Biofuel and Biorefinery Technologies

Volume 7

Series editors

Vijai Kumar Gupta, Department of Chemistry and Biotechnology, Tallinn University of Technology, Tallinn, Estonia
Maria G. Tuohy, School of Natural Sciences, National University of Ireland Galway, Galway, Ireland

This book series provides detailed information on recent developments in biofuels & bioenergy and related research. The individual volumes highlight all relevant biofuel production technologies and integrated biorefinery methods, describing the merits and shortcomings of each, including cost-efficiency. All volumes are written and edited by international experts, academics and researchers in the respective research areas.

Biofuel and Biorefinery Technologies will appeal to researchers and postgraduates in the fields of biofuels & bioenergy technology and applications, offering not only an overview of these specific fields of research, but also a wealth of detailed information.

More information about this series at http://www.springer.com/series/11833

Neha Srivastava · Manish Srivastava
P. K. Mishra · S. N. Upadhyay
Pramod W. Ramteke · Vijaj Kumar Gupta
Editors

Sustainable Approaches for Biofuels Production Technologies

From Current Status to Practical Implementation

Springer

Editors
Neha Srivastava
Department of Chemical Engineering and Technology
Indian Institute of Technology (BHU) Varanasi
Varanasi, Uttar Pradesh, India

Manish Srivastava
Department of Physics and Astrophysics
University of Delhi
New Delhi, India

P. K. Mishra
Department of Chemical Engineering and Technology
Indian Institute of Technology (BHU) Varanasi
Varanasi, Uttar Pradesh, India

S. N. Upadhyay
Department of Chemical Engineering and Technology
Indian Institute of Technology (BHU) Varanasi
Varanasi, Uttar Pradesh, India

Pramod W. Ramteke
Department of Biological Sciences
Sam Higginbottom Institute of Agriculture, Technology and Sciences
Allahabad, Uttar Pradesh, India

Vijaj Kumar Gupta
Department of Chemistry and Biotechnology, School of Science
Tallinn University of Technology
Tallinn, Estonia

ISSN 2363-7609 ISSN 2363-7617 (electronic)
Biofuel and Biorefinery Technologies
ISBN 978-3-030-06916-2 ISBN 978-3-319-94797-6 (eBook)
https://doi.org/10.1007/978-3-319-94797-6

Softcover re-print of the Hardcover 1st edition 2019

Printed on acid-free paper

This Springer imprint is published by the registered company Springer Nature Switzerland AG
The registered company address is: Gewerbestrasse 11, 6330 Cham, Switzerland

Foreword

Biofuels are the potential and sustainable alternative sources of fossil fuels. Efforts are continuously being made to develop economically competitive biofuels and bioenergy. Despite having tremendous efforts, green energy/fuels option did not the make the substantial move and still far from practical implementation globally. To develop economic and viable biofuels, all processing aspects of biomass conversion including mapping, logistic, transportation and storage of feedstock should also be given high priority. Research at laboratory level should be done focusing on the industrial parameters which directly influence the biofuels yield and productivity. Selection of right feedstock and its handling at shop floor, and mechanization of biomass pretreatment need to be taken into consideration. Biofuels production process should be robust and consolidated having minimum processing steps with zero waste discharge. Publication of the book on '*Sustainable Approaches for Biofuels Production Technologies: From Current Status to Practical Implementation*' is a timely and good effort in this direction. This book consists of 10 specific chapters focusing on different kinds of biofuels, existing technologies and sustainable approaches to improve biofuels production process. This book is directed towards presenting drawbacks in existing process and technologies in current biofuels options. The book comprehensively presents the kind of available biofuels options and analyses their potential for using as an alternative to conventional fossil fuel. I am sure this book will serve as one of the key collections of information for the scientists, researchers, teachers and students working in area of biofuels research and development.

I congratulate Dr. Neha Srivastava from IIT-BHU, Varanasi, Dr. Manish Srivastava from DU, Delhi, Prof. (Dr.) P. K. Mishra from IIT-BHU, Varanasi, Prof. (Dr.) S. N. Upadhyay from IIT-BHU, Varanasi, Prof. (Dr.) Pramod W. Ramteke from SHUATS, Allahabad, and Dr. Vijai Gupta from Tallinn University of Technology, Estonia for bringing out this valuable publication on '*Sustainable Approaches for Biofuels Production Technologies: From Current Status to Practical Implementation*' to satisfy the current demand of industries, scientists, teachers, researchers and students.

My sincere thanks go to the editors for their hard work and dedication in this attempt. All the authors and editors of the Series of Biofuels and Biorefinery Technologies deserve sincere appreciation for their efforts in preparing this valuable publication.

Sao Paulo, Brazil

Anuj K. Chandel, Ph.D, M.B.A
Department of Biotechnology
Engineering School of Lorena (EEL)
University of Sao Paulo (USP)

Acknowledgements

The editors are grateful to all the academicians and scientists who have made their effortful contribution to complete this book. We all editors also express our gratitude towards our parents whose continuous support and blessings have always encouraged us to pursue academic activities. It is fairly possible that while completing this task, some mistakes might have snuck in text unintentionally and for these we owe unadulterated responsibility. We are grateful to all authors for their contribution to present book. We are also thankful to our institution/university IIT (BHU) Varanasi, DU, SHUATS and Tallinn University of Technology for giving this opportunity to facilitate this work. We thank them from the core of our heart.

Contents

About the Editors

Dr. Neha Srivastava Institute Post-Doctoral Fellow, IIT (BHU) Varanasi-221005, U.P., India; Mobile no.: +91-9988062681, E-mail: sri.neha10may@gmail.com.

Field of expertise: Biofuels production, microbial bioprocessing and enzyme technologies.

She is currently working as post-doctorate fellow in the Department of Chemical Engineering and Technology, IIT (BHU) Varanasi, India. She has published 23 research articles in the peer-reviewed journals and has three patents. She has been completed her Ph.D. from the Department of Molecular and Cellular Engineering, SHIATS, India in 2016 in the area of bioenergy. She has been received 06 Young Scientist Awards. Presently, she is working on biofuels production (Cellulase enzymes; production and enhancement, biohydrogen production from waste biomass and bioethanol production).

Dr. Manish Srivastava DST-INSPIRE Faculty Fellow, Department of Physics and Astrophysicsm, University of Delhi, 110007, India; Mobile no.: + 91 - 7503757601, E-mail: 84.srivastava@gmail.com, manish_mani84@rediffmail.com.

Field of expertise: Synthesis of nanomaterials and their application as catalyst for development of electrode materials in energy storage, biosensors and biofuels productions.

He is working as DST-INSPIRE faculty in the Department of Physics and Astrophysics, University of Delhi, India. He has published 39 research articles in the peer-reviewed journals, authored several book chapters and filed 1 patent. He worked as a post-doctorate fellow in the Department of BIN Fusion Technology, Chonbuk National University from August 2012 to August 2013. He was an Assistant Professor in the Department of Physics, DIT School of Engineering, Greater Noida, from July 2011 to July 2012. He received his Ph.D. in Physics from the Motilal Nehru National Institute of Technology, Allahabad, India in 2011. Presently, he is working on the synthesis of graphene-based metal oxide hybrids and their applications as catalysts. His area of interest is synthesis of nanostructured materials and their applications as catalyst for the development of electrode materials in energy storage, biosensors and biofuels production.

Prof. P. K. Mishra IIT (BHU), Varanasi-221005, India; Mobile no.: 9415301462, E-mail: pkmishra.che@itbhu.ac.in.

Field of expertise: Biofuels production, microbial bioprocessing and enzyme technologies.

He is currently Professor and Head in the Department of Chemical Engineering and Technology, Indian Institute of Technology (BHU), Varanasi, India. He obtained his Ph.D. degree in Chemical Engineering from Institute of Technology (Banaras Hindu University) in 1995. He has authored/co-authored over 60 technical papers published in reputed national/international journals and supervised more than 20 doctoral students. He has received several awards and honours and has five patents. He is Fellow of Institution of Engineers India. He has received several awards and honours at National/International levels. He has also made significant contribution towards the development of entrepreneurship ecosystem in Eastern part of the country. He is coordinator of Technology Business Incubator at the Institute and member of Executive committee NISBUD, Ministry of Skill Development, Government of India.

Prof. S. N. Upadhyay Emeritus Professor, Department of Chemical Engineering and Technology, Indian Institute of Technology (BHU)-221005, Varanasi, India; Mobile no.: 9415372465, E-mail: snupadhyay.che@itbhu.ac.in.

Field of expertise: Biofuels production, microbial fuel cell, membrane technology and wastewater treatment.

He is an Emeritus Professor in the Department of Chemical Engineering and Technology, Indian Institute of Technology (BHU), Varanasi, India. He has made significant contributions to the institute as professor for 50 years and also served as the institute's director for the academic year (2005–2010). He obtained his Ph.D. (Chemical Engineering) degree from Banaras Hindu University in 1969. He has authored/co-authored over 180 technical papers published in reputed international journals and supervised more than 30 doctoral students in the field of Biochemical and Chemical Engineering. He is Fellow of National Academy of Sciences India, Indian National Academy of Engineering, Biotechnology Research Society, India and Institutions of Engineers, India. He has received several awards and honours at National/International levels. He has made significant contributions to the field of energy/engineering including bioenergy.

Prof. Pramod W. Ramteke Dean of Biological Sciences, SHUATS, Allahabad, U.P., India; Mobile no.: 9415124985; E-mail: pwramteke@gmail.com.

Field of expertise: Biofuels production, plant biology and microbial technology.

He is presently Dean of Post Graduate Studies and Head of the Department of Biological Sciences and Genetics and Plant Breeding at Sam Higginbottom University of Agriculture, Technology and Sciences (SHUATS), Allahabad, India.

He is elected Fellow of National Academy of Agricultural Sciences, National Academy of Biological Sciences, The Linnean Society of London, Academy of Microbial Sciences India (AMI), The Biotech Research Society of India, Society for Plant Protection Sciences, Mycological Society of India, Uttar Pradesh Academy of Agricultural Sciences, Maharashtra Academy of

Sciences, Association for Biotechnology and Pharmacy, Society of Biological Sciences and Rural Development and The Academy of Environmental Biology.

He is recipient of J. C. Bose Gold Medal, SCON Memorial Award for Excellence in Science, Dr. J. C. Edward Medal, Prof. K. S. Bilgrami Memorial Award, Er. V. S. Chauhan Gold Medal, Biotechnology Overseas Associateship and International Fellowship by Biotechnology and Biology Research Council, UK. He is member of editorial board of several journals. He has been a visiting scientist to Hacettepe University, Turkey; Institute of Food Research, UK; University of Liege, Belgium; Korea Institute of Science and Technology, South Korea, University of Szeged, Hungary and Hydrobiological Institute, Czech Republic.

In addition to three patents, three books and over 150 research papers are to his credit. He has implemented seven major projects from DBT, DST, CSIR, UPCAR and CST, UP and supervised 21 Ph.D.s and more than 40 PG Thesis/projects. He was a member of 18th Indian Scientific Expedition to Antarctica (1998–1999).

Dr. Vijai Kumar Gupta ERA Chair of Green Chemistry, Department of Chemistry and Biotechnology, School of Science, Tallinn University of Technology, Akadeemia tee 15, 12618 Tallinn, Estonia; Phone: +372 620 2833, Mobile no.: +372 5671 1014, E-mail: vijai.gupta@ttu.ee; vijaifzd@gmail.com.

He from ERA Chair of Green Chemistry, Department of Chemistry and Biotechnology, School of Science, Tallinn University of Technology, Tallinn, Estonia, is one of the leading experts in the area of microbial biology and biotechnology. He is the member of International Sub-commission on Trichoderma and Hypocrea, Austria; International Society for Fungal Conservation, UK; and Secretary of European Mycological Association. He is the Fellow of Prestigious—The Linnaean Society, London, UK; Fellow of Indian Mycological Association; and Fellow of Mycological Society of India. He has been honoured with several awards in his career including Indian Young Scientist Award for his advanced research achievements in the field of fungal biology and biotechnology. He is the editor of few leading scientific journals of high repute and having many publications in his hands with h-index 21. He has edited many books for publishers of international renown such as CRC Press, Taylor and Francis, USA; Springer, USA; Elsevier Press, The Netherlands; Nova Science Publisher, USA; De Gruyter, Germany; and CABI, UK.

Bioprocessing Perspective in Biorefineries

Sheelendra M. Bhatt and Jatinder Singh Bal

Abstract In the current chapter, the various strategies for biofuel production focusing on rice straw have been discussed. The basic aim is to address the technical applications in enhancing biofuel production using lignocellulosic biomass. The overall price can be minimized using lignocellulosic biomass fractionation at biological platform, where lignin separation and also conversion of biomass into biofuel production is possible at a single platform. The inclusion of chemical pretreatment methods often produces toxic components thus inhibiting the cell to grow during further conversion, thus results in low productivity. In such case, microbial consortia may be a good option. Understanding of bioprocessing steps can lead to the development of sustainable technology for pilot-scale economical productions, to meet the current demand for biofuel.

Keywords Bioprocessing · Microbial consortia · Xylanase · Pectinase
Cellulase · Biorefineries · Biofuel

1 Introduction

In the current scenario, the principal requirements of human being are food feed and fuel. Regarding fossil fuel its going to deplete very soon in coming years. Biofuel is the only alternative to rapidly depleting fossil fuel, because of its sustainability, efficiency, and economics when used for blending thus helps in reducing overall cost. A variety of biofuels exist today of which bioethanol always being in great demand and thus high production is required.

According to a report of 2014, around 74% of biofuel ethanol was produced, while biodiesel is produced in second highest amount (Gupta and Verma 2015). In 2016, the growth rate of global biofuel production was 2.6%, while in current scenario biofuel demand is increasing at 6.5% per annum, while petroleum reserve is decreasing

S. M. Bhatt (✉) · J. S. Bal
CSE Department, Sant Baba Bhag Singh University, Jalandhar, Punjab, India
e-mail: drsmbhatt@gmail.com

N. Srivastava et al. (eds.), *Sustainable Approaches for Biofuels Production Technologies*, Biofuel and Biorefinery Technologies 7,
https://doi.org/10.1007/978-3-319-94797-6_1

Table 1 Composition of biomass

Lignocellulosic biomass	Cellulose	Hemicellulose	Lignin	References
Sugarcane (top)	29.85	18.85	25.69	Sindhu et al. (2011)
Corn stalk	34.45	27.55	21.81	Wu et al. (2011)
Bagasse	30	35	18	Sarkar et al. (2012)
Sugarcane bagasse	44	27	24	de Souza et al. (2013)
Sweet sorghum bagasse	36.9	17.8	19.5	Umagiliyage et al. (2015)
Wheat straw	38.7	19	17.3	Valdez-Vazquez et al. (2015)
Rice straw	35.8	21.5	24.4	Imman et al. (2015)
Rapeseed	51.3	17.3	44	López-Linares et al. (2015)
Corn stover	36.3	31.4	17.2	Saha et al. (2016)

day by day. Of all global productions, India is producing only 1% of biofuel, while consumption rate is high up to 3.1%. Consequently, options are available where many biomasses can be converted into ethanol, biodiesel (Mofijur et al. 2015), and other gaseous or liquid biofuels. There are various advantages of using biofuel such as (1) high energy (2) CO_2 mitigation (3) renewable (4) eco-friendly (5) and can be produced from nonedible biomass.

According to a report of European Union, the target of ethanol production has been set to 8 billion liters up to 2020. Japan has a target of 6 billion liters of ethanol up to 2030. India has a target of 4 billion gallon tons (Energetica India Report 2009). The current costs of biochemical cellulosic ethanol are estimated to be between US$4.03 and $5.60 per US gallon of annual capacity (Binod et al. 2010). The current cost of ethanol today is $1.22 per gallon while it may reduce further up to $70 per gallon (http://www.biofuelsdigest.com/bdigest/2017/05/18/ethanol-and-biodiesel-dropping-below-the-production-cost-of-fossil-fuels/).

The utilization of lignocellulosic biomass includes delignification, hydrolysis, and saccharification before final conversion into ethanol. Complex bonding of lignocellulosic plant structure makes it a recalcitrant owing to the presence of lignin, hemicellulose, and other materials with cellulosic content. Table 1 shows cellulosic composition in various biomasses. Pretreatment is one of the mandatory requirements, in order to get rid of lignin from lignocellulosic biomass. Therefore, delignification is one of the costly and challenging procedures and need technical improvement for pilot-scale applications (Bhatt 2014; Phitsuwan et al. 2013).

In this context, biorefineries concept seems practicable, where every fraction of biomass is processed into value-added product. In the current chapter, we will discuss issues of bioprocessing steps in biofuel ethanol production with a special focus on rice straw which is the globally largest available biomass (Abedinifar et al. 2009).

Sustainable Biofuel Production: Indian Scenario

As per as the report of NUiCONE proceedings, India has now changed strategies to enhance its biofuel production beyond 1%. Therefore, around 85% increase in

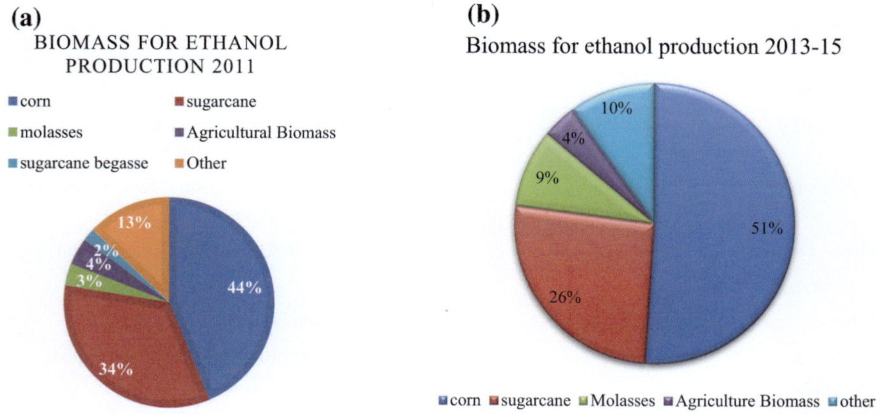

Fig. 1 **a** Bioethanol production from 2013 to 2015. *Source* Araújo et al. (2017), Sindhu et al. (2017). **b** Major biomass for bioethanol production. *Source* Mood et al. (2013), Balan (2014)

biofuel production has been observed since 2009. Also, the Government of India has now changed the blending mandate from 5 to 20% in 2017. Therefore, sustainable technology is required for the enhancing current production of 52.32–83.58 Mt (around >26%). In 2008, the US participation was around 51%, Brazil was 36%. EU's contribution was only 4% to ethanol production, while in 2011, US contribution was around 46%, Brazil was 22%, while EU was 17% see Fig. 1.

The United States is the world's largest producer of ethanol, and has produced nearly 15 billion gallons in 2015 alone. Together, the U.S. and Brazil produce 85% of the world's ethanol. The vast majority of U.S. ethanol is produced from corn, while Brazil primarily uses sugar.

2 Biofuel Classifications

Based on the types of biomass, biomass can be classified as mentioned in Table 2 and various feedstock used in biofuel production has also been mentioned. Out of which corn and molasses are used mainly for biofuel ethanol production. Biodiesel production is another strategy to cope up with demand in the current scenario, which uses various agricultural and nonagricultural waste as a source. Ethanol production has been mentioned in billions of gallons of different biomasses.

First generation (1st G) biofuel includes sugar crops (sugar beet, sugarcane), edible crops (corn, sorghum), oilseed crops (soybean, canola), and animal fats. Second generation (2nd G) biofuel includes all types of cellulosic biomass, nonfood crops, and waste biomass, while third (3rd G) generation biofuel is based on the use of algae and municipal solid wastes (Singh et al. 2011).

Table 2 Biofuel classification

Fuel	Feedstock
First generation	
Bioalcohol ethanol, Propanol butanol	Starches from wheat, corn, sugar cane, molasses, potatoes, other fruits
Biodiesel	Oils and fats including animal fats, vegetable oils, nut oils, hemp, and algae
Green Diesel	Made from hydrocracking oil and fat feedstock
Second generation	
Ethanol	Lignocellulosic biomass straw, bagasse
Biobutanol	Lignocellulosic biomass straw, bagasse
Third generation	
Bio-oil	Algae
Fourth generation	
Photobiological solar fuels and electrofuels	

There are various generations of biofuel such as first, second, third, and fourth generation biofuel. First generation biofuel includes all kind of edible feedstock which can be hydrolyzed by simple steam and further by enzymatic conversion into ethanol. Second generation biofuel is mostly based on the use of lignocellulosic biomass for biofuel production. These are less competitive to edible crops and are mostly available as agriculture, municipal, or industrial waste (Araújo et al. 2017). This biomass is rich in cellulose and hemicellulose which can be converted into ethanol after several bioprocessing steps. Recent research shows that production of bioethanol from these wastes is little expansive due to (1) feedstock cost, (2) feedstock harvesting, (3) feedstock densification, (4) feedstock pretreatment, (5) by-product separations (6) environmental and health impact. Biobutanol is another fuel which is obtained using the same feedstock but with different microbes (Araújo et al. 2017). Third-generation biofuel is obtained as bio-oil using algae as feedstock while fourth-generation biofuel is obtained as bio-solar fuel and electro-fuel as shown in Table 2.

Economically sustainable lignocellulosic-based ethanol production depends on

(1) Suitable substrate
(2) Suitable pretreatment applications without inhibitors productions
(3) Suitable biocatalyst
(4) Robust yeast cell bioethanol conversion (e) proper detoxification.

The order of cost of production is 3G > 2G > 1G. The ethanol production cost of molasses-based feedstock varies between 0.78 and 0.97 US$/L. However, production of ethanol at pilot scale is still in process of demonstration scale from the 3G feedstock. The cheapest ethanol production is in Brazil, where a combination of readily available resources and cheap labor makes prices of about $0.20 per liter possible.

3 Lignocellulosic Ethanol: Production Technology

There are various modes of ethanol production using rice straw such as SSF, SHF, sequential hydrolysis and fermentation process, SSMSF, SESF (Abedinifar et al. 2009; Karimi et al. 2006; Shinozaki and Kitamoto 2011; Ko et al. 2009). Each technique has its own merits and demerits but the basic idea is to utilize the pentose sugar present in hemicellulose. Thus, much novel work has been reported such as the use of lactic acid bacteria which has the capability to produce lactic acid during fermentation and helpful in the insolubilization of cellulose that is able to ferment glucose and (Kim et al. 2010). Thus, it was an integration of enzymatic hydrolysis and fermentation in one step called separate hydrolysis and fermentation. Ethanol production by *Mucor indicus* and *Rhizopus oryzae* from rice straw was successful in bypassing the end product inhibition (Abedinifar et al. 2009).

The bioconversion of cellulosic materials includes the formation of soluble sugars from cellulose in paper/agricultural residues and depends on the coordinated action of individual components such as β-exoglucanase, β-endoglucanase, β-glucosidase of cellulase enzyme.

3.1 Ethanol from Lignocellulosic Biomass

There are various biomasses used for lignocellulosic-based ethanol production such as rice and paddy straw because, water hyacinth corn stover. Predominantly, bioethanol is derived from stover and switchgrass in the U.S., while sugarcane bagasse is used mainly in Brazil and India, and rice husk and straw from China and India (Khoo 2015).

3.1.1 Corn Stover

Ethanol production from corn stover (Saha et al. 2013) was studied by many workers. Corn stover is rich in cellulose, i.e., 37% while having less lignin comparatively. Hydrothermal pretreatment combined with enzymatic saccharification leads to conversion of 72% glucose (Saha et al. 2013) while ethanol yield was 0.49 g/g biomass. Corncob was reported to produce more furfural as compared to corn stover, which is a toxic component released after pretreatment and should be an inhibitor for the growth of microbes involved in hydrolysis and saccharification. Therefore, corn stover is suitable for ethanol production (Kadam and McMillan 2003). Microbial pretreatment of corn stover with *Ceriporiopsis subvermispora* was attempted with enzymatic hydrolysis by some workers, which resulted in 66% production of glucose in 35 days (Wan and Li 2010). In another work, steam pretreatment was given in SSF mode which resulted in 70% ethanol with 10% insoluble solid biomass with yeast concentration 5 g/l. In SSF mode, alkaline pretreatment of corn stover coupled with fungal treatment with *Phanerochaete chrysosporium or Gloeophyllum trabeum* and fermentation with *Saccharomyces cerevisiae or Escherichia coli K011* resulted in 3.09. g/100. g stover at day 4 (Vincent et al. 2014). In an attempt to resolve the problem of released toxic during pretreatments such as furfural and other inhibitory compounds, corn stover were hydrolyzed and hemicellulose fraction was treated with by roto-evaporation and lime neutralization resulted in removal of more than 50% furfural and acetic acid, which was volatile and resulted in enhanced ethanol production of 31.1 g/L and the corresponding ethanol yield on fermentable sugars of 0.406 g/g were obtained within 72 h in batch fermentation of the detoxified hydrolysate with immobilized cells (Zhao and Xia 2010). In another report, corn stover was pretreated with dilute H_3PO_4 (0.0–2.0%, v/v) or 1% acid combined with biological treatments with Escherichia coli strain FBR and condition were optimized by response surface methodology resulted in 85% glucose yield. In a similar effort to produce butanol using ABE technology, simultaneous saccharification, fermentation, and recovery (SSFR) were attempted which resulted in hydrolysis of 97% sugar from corn stover (Qureshi et al. 2014).

3.1.2 Rice Straw

Rice straw burning was recently in the news in Punjab at a large level, and as a proof images were released as shown in Fig 2. As an estimate, around 2.5 tons of rice straw has been burnt per acre of land due to the season of next crop within 20 days. In Punjab, around 11 MT (million tons) of rice is grown which produces around 21 MT of rice straw alone which resulted in the release of >70% CO_2 7% CO, and 0.66% CH_4 along with 2.09% N_2O (Binod et al. 2010). Besides warning by the government and court, farmers are burning this useful biomass due to the lack of appropriate technology. One such attempt has been done to convert them into pellets which later can be used into fuel generations by using techniques of pyrolysis. The need is to develop farmer-friendly technology, which can be used in one step to yield glucose and thus ethanol.

4 Bioprocessing of Lignocellulosic Biomass

Removal of recalcitrant lignin requires various complex chemical or biochemical processing to get rid of cellulose and hemicellulose, which is the main carbohydrate to be used in fermentation and sachharification.

Different studies show that composition of biomass depends on various biotic and abiotic factors, ecosystems and time of harvesting. Variation of lignin content varies from 3 to 20%, cellulose content from 17 to 14%, which is the most suitable for ethanol production. After separation of cellulosic biomass, enzymatic hydrolysis and fermentation lead to the conversion of ethanol production. Owing to environmental issue combination of treatment is preferred to make biochemical process feasible in

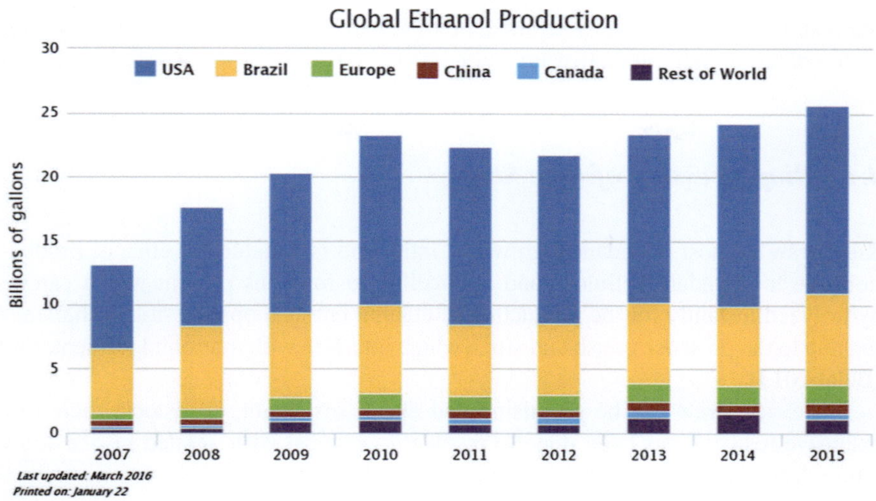

Fig. 2 *Source* RFA renewable fuel associations https://www.afdc.energy.gov/data/10331

one step. A recent review of Sindhu et al (2016) shows that combined pretreatment is more efficient in delignification as compared to a single chemical pretreatment process (Sindhu et al. 2016). Mood et al. (2013) show that alkali pretreatment alone can be used with other pretreatment processes even with enzymatic pretreatment (Mood et al. 2013), which is ineffective and time consuming, therefore, Mishima et al (2008) show that enzymatic efficiency can be improved by using at least 20 chemical pretreatment methods (Mishima et al. 2008; Klein et al. 2016). According to Singh et al. (2015), lignocellulosic biomass pretreatment is challenging and need further research for making it cost-effective (Singh et al. 2015). As per our own research Bhatt (2014), alkali pretreatment leads to less solubilization of cellulose and hemicellulose than acid pretreatment (Bhatt 2014). To improve enzymatic treatment, other strategies have been worked out by some workers such as the use of ionic liquids, or use of microwave-assisted technology. In such a case, the main objective was to reduce the solubilization of cellulose and hemicellulose while maximizing the removal of Lignin. In this regard, the experiment of Klein et al. (2016) demonstrated that microwave-assisted chemical pretreatment is more effective for enzymatic-based lignin removal (De Bhowmick et al. 2017; Klein et al. 2016). The main aim is to save the environment by using fewer chemicals during pretreatment.

Therefore, fermentable sugar obtained after enzymatic hydrolysis is more effective for ethanol production. A similar work of Xia et al. (2013) shows that up to 94.6% sugar can be derived by using acid pretreatment (Kapoor et al. 2017; Xia et al. 2013).

Ethanol production from rice straw uses either solid-state fermentation (SSF) or simultaneous saccharification and fermentation (SSF) or by separate enzymatic hydrolysis and fermentation (SHF) (Singh et al. 2016). SHF may be a better option as compared to SSF mode (Akhtar et al. 2017). In some experiments, microwave pretreatment was combined with alkali pretreatment, and SSF mode was more useful as compared to SHF mode (Swain and Krishnan 2015).

4.1 Bioprocessing of Rice Straw

Rice straw is most abundantly grown in India and is suitable for ethanol production due to abundant cellulose and hemicellulose fractions present which can be hydrolyzed instantly for the production of ethanol but its bioprocessing is challenging due to the presence of ash and silica which interfere with microbial fermentation (Belal 2013).

Since rice straw can be promising and also a sustainable feedstock, therefore, extensive research has been done to resolve a technical issue related to rice straw biomass conversion a disadvantage linked with this biomass.

Bioprocessing of rice straw can be divided into four main parts (1) pretreatment which includes chemical, physical, and biological or combined, (2) mode of saccharification and fermentation for ethanol productions. Therefore, the choice of suitable pretreatment technology is a deciding factor where a large fraction of cellulose and hemicellulose can be available for hydrolysis and enzymatic saccharification. In one report, pretreatment of various lignocellulosic biomass including rice straw with *Trichoderma reesei* resulted in enhanced hydrolysis of rice straw comparatively (Singh et al. 2016) (Fig. 3).

4.1.1 Pretreatment of Rice Straw

The main objective of pretreatment is to reduce crystallinity and degree of polymerization of rice straw so that potential carbon can be unlocked such as cellulose, hemicellulose, and other components such as lignin. In this regard steaming, milling, irradiation, and temperature along with pressure are applied which is useful in reducing crystallinity (Nguyen et al. 2010; Poornejad et al. 2013; Chang et al. 2016, 2017). The purpose of milling is to reduce the surface area of rice stalk and making available of rice husk biomass for biofuel production. Since powdered form of rice straw is fine and can be converted into animal feedstock but due to the presence of large amount of silica, RICE straw cannot even be converted into animal feed since presence of silica can tear the jaw and can give wounds in animals, therefore getting rid of silica is not an easy task.

The only drawback of rice straw is the presence of around 75% silica while its advantages are its availability in large amounts in the world. Presence of Silica makes bioprocessing of rice straw tough, thus, it is even composting is difficult. Presence of silica acts as a barrier in the protection of leaves and is present over leaves and also as a layer of plant part thus, its hydrolysis is also very difficult. Only a few reports are available around the world for technology which can make rice straw free of silica.

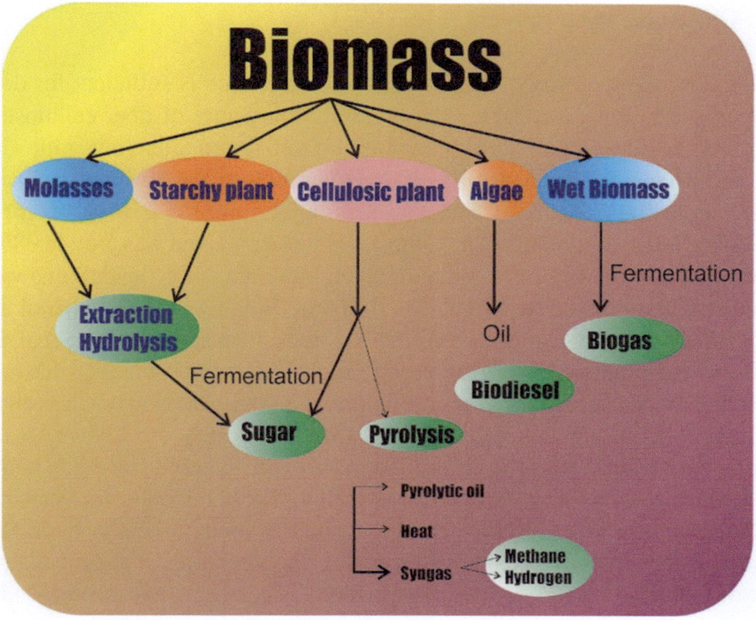

Fig. 3 Bioprocessing of ethanol production from lignocellulosic biomass. *Source* https://mediawiki.middlebury.edu/wiki/OpenSourceLearning/Biofuels, http://biofuel.org.uk/types-of-biofuels.html

Some workers had attempted some new methodologies to get rid of this silica from rice husk (Araújo et al. 2017). Its composting is also very difficult owing to the presence of silica and lignin. Also, treatment with ionic liquids is helpful for complete lignin removal. Various ionic liquids have experimented till date (Sindhu et al. 2017). Some workers reported that in presence of SDS, ionic liquid is able to release lignin up to 46% by increasing the temperature up to 100 °C Lau et al. (2015) along with tetrabutyl-phosphonium hydroxide resulted in the removal of silica before lignin removal (Lau et al. 2015). One more recent work shows that silica can be removed up to 91% with combined pretreatment of organosolv and with sodium carbonate (Khaleghian et al. 2017).

4.1.2 Mechanical Pretreatment

Milling is beneficial in the reduction of particle size which is advantageous in increasing surface/volume ratio of straw which makes chemical or biological treatment accessible. Milling includes chopping, or grinding or pressing.

4.1.3 Ultrasonic Pretreatment

Ultrasound releases approximately 10–100 kJ/mol, which is sufficient for decreasing the crystallinity by destroying the microfibril structure of fiber cellulose up to 78.4–66.3%, and mean particle size reduced up to 0.4 mm after sonication (Bussemaker and Zhang 2013). Generally, 40 kHz is employed at an industrial scale of 40 kHz, about 0.025 W/mL, 25 °C, at 30 min, (Luo et al. 2014; Chuetor et al. 2015). With fungal treatment at 28 °C the net glucose obtained was around 38% with rice straw. Mostly acidic pretreatment or alkali pretreatment is done. Some workers reported that upon mixed treatment of 0.5 M NaOH and with 60% ethanol around 100% removal of lignin was obtained when treated on wheat straw (Sun et al. 2016). Irradiation is helpful in breaking the strong bond which is almost impossible to break by any other mode but certainly, they are helpful with other methods such as chemical or heating.

Strong cavitational effect of ultrasound mixed with suitable solvent is helpful in reducing the load of lignin complexed with cellulose or hemicellulose. Many reports show the combined effect of ultrasound and alkaline pretreatment or hydrothermal treatment is very effective in other lignocellulosic biomasses such as bagasse and rice straw (Wu et al. 2017). Since the technology is costly so, not recommended for industrial applications.

4.1.4 Microwave Treatment

Some reports show the use of microwave along with alkali pretreatment and shows to increase enzymatic digestibility of rice straw (Singh et al. 2014) and microwave pretreatment with organic solvent (acetic acid and propionic acid) further reported to increase enzymatic digestibility of rice straw as a result glucose yield was obtained up to 80% (Gong et al. 2010).

4.1.5 Chemical Pretreatment

Rice straw and rice stalk have been pretreated with many methods such as aqueous ammonia (Araújo et al. 2017), dilute acid (Lee et al. 2015), sodium carbonate and fungus *Mucor hiemalis*, (Khaleghian et al. 2015), alkaline pulping and steam explosion pretreatment (Ibrahim et al. 2011), calcium capturing by carbonation (Park et al. 2010), microwave alkali pretreatment (Singh et al. 2011), dilute sulfuric acid and sulpho-methylation (Zhu et al. 2015), using a cocktail of hydrolytic and oxidizing enzyme (Dhiman et al. 2015), organic acid treatment (Amnuaycheewa et al. 2016), biological pretreatment (Bak et al. 2009; Salvachúa et al. 2011; Okamoto et al. 2011; Arora et al. 2016; Karimi et al. 2006; Bak et al. 2010; Das et al. 2013).

Though many pretreatment technologies are available as discussed previously but every technology has its own advantages and disadvantages, for example, acid treatment has the benefit that it can help in the dissolution of lignin from cellulose before hydrolysis (Mishima et al. 2008; Khaleghian et al. 2017). Though it needs a large quantity of acids which is not environmentally friendly but dilute acid has advantages that it can protect the conversion of hemicellulose into xylan and other inhibitory compounds such as furfural which acts as an inhibitor for the action of cellulase and other microbial enzymes. Acid pretreatment sometimes require high temperature up to 180 °C and use of organic acid is helpful in increasing cellulose depolymerization, for example, pretreatment of the rice straw with 75% (v/v) aqueous ethanol and 1% w/w H_2SO_4 at 150 °C for 60 min resulted in the production of total sugar concentration up to 31. g/L (Amiri et al. 2014). Many workers have got good results with the use of concentrated phosphoric acid during pretreatments of rice straw (Moradi et al. 2013) and reported that now enzymes are more accessible and thus enhance enzymatic loading reported (Amiri et al. 2014). Further, lowering of acid concentration of acid H_2SO_4 (0.25%v/v), HCl, H_3PO_4, and oxalic acid and NaOH (0.25w/v) helps in more release of glucose yield up to 84–91% with very low concentration of furan from rice straw. Pretreated rice straw with acid hydrolysate technology uses the term PRSAH, where rice straw is treated with 1% acid and 1% alkali is beneficial in increasing cellulose content from 38 to 50% during enzymatic treatment with glucose yield 0.58 g/g from PRASH (Chen et al. 2014).

Though dilute acid pretreatment is useful in the rapid hydrolysis of hemicellulose and also in the release of cellulose providing a path for good enzyme accessibility, but they also reported to have many disadvantages such as formation of various inhibitors (Wi et al. 2013; Balan 2014; De Bhowmick et al. 2017). Acid treatment converts glucose into HMF, while xylose into furfural, along with acetic acid and formic acid; or may convert lignin into derived phenolics, oligomers, and re-polymerized furans or pseudo-lignin (Jönsson and Martín 2016). Some authors agree that before acid treatment there should be alkali pretreatment, which helps in good release of hemicellulose.

4.1.6 Alkaline Pretreatment

Use of alkali pretreatment is for breaking lignin from recalcitrant cellulose and hemicellulose. Alkali is useful in breaking the ester bond present in lignin and hemicellulose. A little supply of high temperature is helpful in breaking the ether bond. Thus, the overall effort is very high solubilization of hemicellulose and lignin. Aqueous ammonia and alkaline peroxides are helpful in overall increase impact in promoting solubilization (Cabrera et al. 2014). Alkaline peroxide pretreatment is mostly helpful in reducing the temperature requirement for enhanced saccharification by enzymatic hydrolysis at 30 °C with reducing sugar up to 92% using rice hull. A comparison of inhibitors released during acid or alkali pretreatment shows that most of the inhibitors such as formic and acetic acids and phenolic compounds, while 5-hydroxymethylfurfural (HMF) and furfural are released during acid pretreatment and while during alkali pretreatment inhibitors were not released (Bolado-Rodríguez et al. 2016). An increase of temperature, reduction in time of incubation of enzyme was observed for 60% lignin removal along with an increase in crystalline index from

40 to 52, and in SSF mode there was 98% of conversion yield of ethanol using rice straw (Phitsuwan et al. 2017). In another work, the use of sodium carbonate at mild conditions followed by fermentation using Zygomycetes fungus *Mucor hiemalis* (Khaleghian et al. 2015) results in 90% removal of silica from rice straw while 65% increased enzymatic hydrolysis in SSF mode at 100 °C.

A novel pretreatment study was conducted by Silva et al. (2013), where the objective was to detoxify the inhibitors produced during acid pretreatment by the use of Ozonation in alkaline medium (pH8) in the presence of H_2O_2 and ethanol production was done by using *Pichia stipius* yeast (Silva et al. 2013).

4.1.7 Biological Pretreatment of Rice Straw

The biological/enzymatic application has less been in use for ethanol production at industrial scale. There are many challenges which need attention. Pretreatment is done usually to separate unwanted lignin complex with cellulose and hemicellulose and another problem with rice straw is the presence of silica (15%). Silica separation has not been attempted using any of biological pretreatment methods. Silica and lignin reduce the overall activity of microbes. Microbial growth ceases due to release of inhibitors by lignin, cellulose, or hemicellulose degradation at high acidic pretreatment and high temperature. Laccase is one of the microbes which is known to degrade and solubilize lignin from lignocellulosic biomass. Method for separation of silica was suggested by Ludueña et al. (2011) in which overnight soaking with KOH was suggested in 1:12 ratio (Ludueña et al. 2011).

Laccase is copper-containing oxidative enzyme, which is produced by fungi from class Basidiomycetes, ascomycetes, and deuteromycetes in solid-state fermentation mode.

Under optimal conditions of environmental factor, microbes start producing laccase, lignin peroxidase, and phenoloxidase. Some wood-rotting fungi called as white rot fungus and some mushrooms such as Pleurotus are also known to produce such enzymes. Wheat and rice straw have been used in the past to produce enzymes such as laccase and cellulase (Lee et al. 2012; Nakanishi et al. 2012; Jin and Ning 2013; Parenti et al. 2013; Rastogi et al. 2016; Postemsky et al. 2017).

At pilot scale, there are only a few reports for biological pretreatment-based ethanol production. There are many reports of application of biological pretreatments (Bak et al. 2010; Toquero and Bolado 2014; Mustafa et al. 2017). Mostly fungus is used for combined pretreatment since treatment alone is not useful (Bak et al. 2009). Several Basidiomycetes species such as *Ceriporiopsis subvermispora, Phanerochaete chrysosporium, Pleurotus ostreatus, Phlebia subserialis, and Pichia guilliermondii* can grow on different lignocellulosic biomass have been evaluated for their delignification efficiencies (Kumar et al. 2009). Now, it has been realized that use of microbial consortia (mix of bacteria and fungus) is more helpful in value-added product formation after pretreatment (Shen et al. 2018; Toquero and Bolado 2014). Generally, biological pretreatment with laccase secretion leads to lignin dissolution but also loss of cellulose leads to overall decrease in ethanol production.

To cope up with this problem, use of NaCl is the suggested and the addition of which can control the growth of cells, this technique is called as inhibitor-mediated-intensified biological pretreatment technology IMBP (Kumar et al. 2017). Kogo et al. (2017) have used *Trichoderma reesei and Humicola insolens* for simultaneous enzyme production and hydrolysis (Kogo et al. 2017). *Trichoderma* and *Humicola* are best to known produce cellulase enzyme, which shows an increased effect in alkaline pretreatment.

4.1.8 Microbes for Pentose Utilizations

Pentose utilization is a major issue in the conversion of lignocellulosic biomass into ethanol. In the past, many microbes were genetically modified to utilize pentose released from rice straw hydrolysate, for example, genetically engineered strain *Corynebacteriu glutamicum* wild type was modified

4.1.9 Application of Microbial Consortia

Microbial consortia are important in the current scenario for biological pretreatment since a group of microbe is more effective as compared to the single microbe. The synergistic action of microbes results in improving enzyme activity thus rapid action is expected. This can solve the most problematic part of lignocellulosic digestion, which is lignin degradation and its degradation takes a number of days. Microbial degradation of lignin using microbial consortia has been described in detail by many researchers around the world but still less adopted by the industries due to various challenges (Ding et al. 2016; Jia et al. 2016; de Lima Brossi et al. 2016). Laccase (EC 1.10.3.2) is known as lignin degrader enzyme (multicopper blue oxidase) that couples the four electron reduction of oxygen with the oxidation of a broad range of organic substrates, including phenols, polyphenols, anilines, and even certain inorganic compounds by a one-electron transfer mechanism (Margot et al. 2013).

Consortia are interactive groupings of microorganisms ranging from defined species communities to undefined, multispecies aggregations. Further, several aspects of applied microbial consortia have been reviewed. Some workers have discussed the advantages of using consortia and the difficulty in achieving selective biofuel production. Microbial consortia have been demonstrated because of their enhanced characteristics over monoculture approaches in the conversion of cellulose and other sugar mixtures to alcohol (Xing et al. 2012). Wan and Li (2010) worked on microbial delignification of corn stover by *Ceriporiopsis subvermispora* for improving cellulose digestibility. MnP and laccase were detected during the degradation of corn stover by *C. subvermispora*. For major hydrolytic enzymes, xylanase was the only enzyme detected which resulted in 39% lignin degradation. Overall glucose yield was about 72% after the enzymatic hydrolysis in 18 days (Wan and Li 2010).

Zuroff and Curtis (2012) reviewed on developing symbiotic consortia for lignocellulosic biofuel production. The author concluded that the designing consortia with an understanding of cooperative microbial energetics could allow the development of efficient biofuel production processes using existing natural or genetically modified/selected organisms and could pave the way for future bioprospecting and genetic engineering. Engineering microbial consortia to produce biofuel involves short-circuiting the catabolic cascade to accumulate the biofuel of interest. The energetic, metabolic, and physiological conditions that allow this to occur are the key process in design considerations for a biofuel production platform. Natural and engineered interactions appear to be promising methods for community control and regulation. In the meantime, ample lignin degrading, cellulolytic, and fuel-producing organisms have been characterized and are available to explore the potential of symbiotic relationships for biofuel production. Decreased rates and the relatively low value of biofuels suggest the need for a new paradigm of low-cost bioprocessing technology. Organism, consortia, and bioprocess design must advance hand-in-hand with technical and economic feasibility in order to make lignocellulosic biofuels a reality (Zuroff and Curtis 2012).

5 Bioreactor and Optimizations Conditions

Bioreactor design at pilot-scale production of ethanol using a variety of cellulosic biomass operates in various modes such as batch (Gusakov et al. 1985), airlift (Zheng et al. 2005), packed bed reactor (Canabarro et al. 2017), batch tube reactor for biomass hydrolysis (Reactor and Engineering 2001), rotating fibrous bed reactor (Lan et al. 2013) for ethanol production. There is a continuous research for improvement process parameters in order to improve the cost of production and accordingly conditions are adjusted. As depicted in Fig. 4, there are four main types of processing commonly operating for industrial ethanol production. (1) Separate Hydrolysis and Fermentation (SHF); (2) Simultaneous Saccharification and Fermentation (SSF); (3) Simultaneous Saccharification and Co-fermentation (SSCF); (4) Consolidated Bioprocessing (CBP).

5.1 Separate Hydrolysis and Fermentation (SHF)

A very good review has been presented by Nguyen et al. (2017) in which pros and cons of SHF has been discussed. SHF process includes enzyme production, hydrolysis, hexose and pentose utilization separately, which has benefits that it can reduce the end product inhibitions but on the other hand reduction in the yield of ethanol has been reported. It may be due to low enzyme loading, culture in stress condition that perform lower production. It also has been reported that in SSF mode carbon starvation takes place before glucose consumptions. Temperature beyond 37 °C lowers ethanol production while at 37 °C ethanol production was around 80 g/l. Also, because of high temperature, low cell viability was observed.

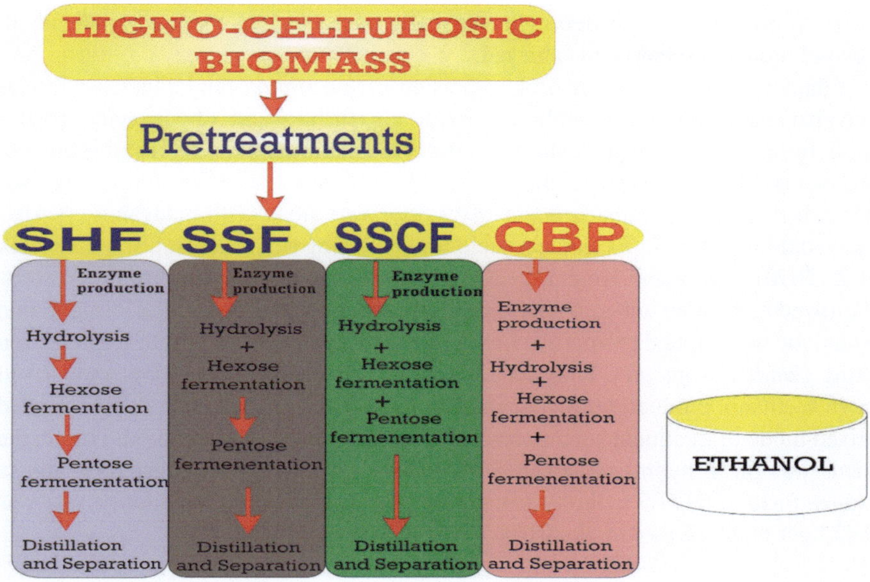

Fig. 4 *Source* Modified from (Devarapalli and Atiyeh 2015) abbreviation used SHF = Separate Hydrolysis and Fermentation; SSF = Simultaneous Sachharification and Fermentation; SSCF = Simultaneous Sachharification and Co-fermentation; CBP = Consolidated Bioprocessing

There is a separate reactor, thus hexose and pentose utilization occurs separately and proceeds at optimum conditions. Many factors related to the production of enzymes and fermentation can be set to optimum in separate hydrolysis and fermentation. The main limitation is the inhibition of enzyme production because of glucose and cellobiose, therefore during hydrolysis overall efficiency is reduced.

5.2 *Solid-State Fermentation (SSF)*

Solid-state fermentation is another condition for production of ethanol from cellulose, where first, enzyme production is done and then enzyme is mixed with cellulosic biomass for hydrolysis. SSF condition requires mixing of microbes to enhance the biomass fermentation simultaneously in order to relieve the product inhibition by the sugar produced during fermentation (Swain and Krishnan 2015; Bak et al. 2010). The main challenges encountered during production is providing an optimal condition for microbes to produce enzymes cellulase.

Therefore, another strategy was adapted as SSCF (see Fig. 4), where both hexose and pentose have been co-fermented by using mostly genetically modified *S. cerevisiae and Zymomonas mobilis*. While in CBP, both hydrolysis and fermentation are performed in a single step by using single microbes thus the high efficiency of ethanol production has been achieved.

Ethanol from rice straw hydrolysate using *Pichia stipitis* can be increased twice and also results in reduced inhibitors production (furfural and 5-hydroxymethyl furfural) by using ammonia. In addition xylose fermentation could result in more ethanol production (Lin et al. 2012). Bak et al. (2010) demonstrated that fungal pretreatment of rice straw (*Phanerochaete chrysosporium*) along with manganese peroxidase could result in 62.7% ethanol yield for 96 h using rice straw in SSF mode (Bak et al. 2010). Swain and Krishnan (2015) demonstrated that xylitol from rice straw improved much after aqueous ammonia pretreatment, where sequential fermentation technique was applied to improve lignin digestibility in repeated batch fermentation using *Candida tropicalis*. Thus adapting two-stage batch fermentation could result in 98% ethanol production (Swain and Krishnan 2015). Zahed et al. (2016) applied mixed mode of treatment such as batch and continuous co-fermentation. Hence, the focus was on to hydrolyze substrate rice straw into maximum sugar 81% and to reduce furfural by 505 xylitol yield was 68% in continuous mode dilution rate was 0.03 l per hours (Zahed et al. 2016).

6 Future Scope

In summary, we can conclude that the future of biofuel lies in the economical production of ethanol using rice straw using microbial consortia-based technology and adapting suitable bioreactor condition platform so that maximum utilization of biomass can be done in an efficient way. As per the report, EU has set a target of 10% bioethanol production but also has to reduce CO_2 emission by 6% against the current emission of more than 30% (Union 2009). Therefore, for sustainable production of ethanol using largest available biomass on earth, rice straw requires adapting continuous bioreactor condition along with batch fermentation which may be helpful in minimizing inhibitor formation after acid pretreatment under co-fermentation conditions (Zahed et al. 2016). As per data available, bioprocessing of 1 ton of rice straw yields around 239–253 L ethanol with 292 kg CO_2 eq/ton straw (Soam et al. 2016).

The author reported that 1% increase in enzyme during hydrolysis increases ethanol yield to 2.9%. The author also reported that reducing chemical pretreatment is essential since it accounts for release of 30% CO_2 while biological pretreatment also account for the release of CO_2 emissions which can be reduced by recycling of enzymes at various stages of hydrolysis of biomass. Against dilute acid treatment, steam explosion technique is a more viable option for large-scale biomass pretreatment. Another benefit of evaluation of techno-economical aspect is reducing rice straw biomass burning, improving socioeconomical aspects of farmers and thus helpful in setting favorable policy as per Indian scenario concerns.

Acknowledgements The author expresses thanks to all faculties of CoD of Natural Science Department, Director R&D department, and special thanks for VC, SBBSU, Director Dr. Vijay Dhir and all those who are involved directly or indirectly in supporting me during manuscript preparations.

Conflict of Interest The author declares no conflict of interest with any agency.

References

Abedinifar S, Karimi K, Khanahmadi M, Taherzadeh MJ (2009) Ethanol production by *Mucor indicus* and *Rhizopus oryzae* from rice straw by separate hydrolysis and fermentation. Biomass Bioenerg 33(5):828–833. https://doi.org/10.1016/j.biombioe.2009.01.003

Akhtar N, Goyal D, Goyal A (2017) Characterization of microwave-alkali-acid pre-treated rice straw for optimization of ethanol production via simultaneous saccharification and fermentation (SSF). Energy Convers Manag 141:133–144. https://doi.org/10.1016/j.enconman.2016.06.081

Amiri H, Karimi K, Zilouei H (2014) Organosolv pretreatment of rice straw for efficient acetone, butanol, and ethanol production. Biores Technol 152:450–456. https://doi.org/10.1016/j.biortech.2013.11.038

Amnuaycheewa P, Hengaroonprasan R, Rattanaporn K, Kirdponpattara S, Cheenkachorn K, Sriariyanun M (2016) Enhancing enzymatic hydrolysis and biogas production from rice straw by pretreatment with organic acids. Ind Crops Prod 87:247–254. https://doi.org/10.1016/j.indcrop.2016.04.069

Araújo K, Mahajan D, Kerr R, da Silva M (2017) Global biofuels at the crossroads: an overview of technical, policy, and investment complexities in the sustainability of biofuel development. Agriculture 7(4):32

Arora A, Priya S, Sharma P, Sharma S, Nain L (2016) Evaluating biological pretreatment as a feasible methodology for ethanol production from paddy straw. Biocatal Agric Biotechnol 8:66–72. https://doi.org/10.1016/j.bcab.2016.08.006

Bak JS, Ko JK, Choi IG, Park YC, Seo JH, Kim KH (2009) Fungal pretreatment of lignocellulose by *Phanerochaete chrysosporium* to produce ethanol from rice straw. Biotechnol Bioeng 104:471–482. https://doi.org/10.1002/bit.22423

Bak JS, Kim MD, Choi IG, Kim KH (2010) Biological Pretreatment of Rice Straw by Fermenting with *Dichomitus squalens*. New Biotechnol 27:424–434. https://doi.org/10.1016/j.nbt.2010.02.021

Balan Venkatesh (2014) Current challenges in commercially producing biofuels from lignocellulosic biomass. Int Sch Res Not 2014(May):e463074. https://doi.org/10.1155/2014/463074

Belal EB (2013) Bioethanol production from rice straw residues. Braz J Microbiol [Publication of the Brazilian Society for Microbiology] 44(1):225–234. https://doi.org/10.1590/S1517-83822013000100033

Bhatt SM (2014) Lignocellulosic feedstock conversion, inhibitor detoxification and cellulosic hydrolysis—a review. Biofuels 5(6):633–649. https://doi.org/10.1080/17597269.2014.1003702

Binod P, Sindhu R, Singhania RR, Vikram S, Devi L, Nagalakshmi S, Kurien N, Sukumaran RK, Pandey A (2010) Bioethanol production from rice straw: an overview. Biores Technol 101:4767–4774. https://doi.org/10.1016/j.biortech.2009.10.079

Bolado-Rodríguez S, Toquero C, Martín-Juárez J, Travaini R, García-Encina PA (2016) Effect of thermal, acid, alkaline and alkaline-peroxide pretreatments on the biochemical methane potential and kinetics of the anaerobic digestion of wheat straw and sugarcane bagasse. Biores Technol 201:182–190. https://doi.org/10.1016/j.biortech.2015.11.047

Bussemaker MJ, Zhang D (2013) Effect of ultrasound on lignocellulosic biomass as a pretreatment for biorefinery and biofuel applications. Ind Eng Chem Res 52 (10):3563–3580

Cabrera E, Muñoz MJ, Martín R, Caro I, Curbelo C, Díaz AB (2014) Alkaline and alkaline peroxide pretreatments at mild temperature to enhance enzymatic hydrolysis of rice hulls and straw. Biores Technol 167:1–7. https://doi.org/10.1016/j.biortech.2014.05.103

Canabarro NI, Alessio C, Foletto EL, Kuhn RC, Priamo WL, Mazutti MA (2017) Ethanol production by solid-state saccharification and fermentation in a packed-bed bioreactor. Renew Energy 102:9–14. https://doi.org/10.1016/j.renene.2016.10.026

Chang KL, Chen XM, Han YJ, Wang XQ, Potprommanee L, Ning XA, Liu JY et al (2016) Synergistic effects of surfactant-assisted ionic liquid pretreatment rice straw. Biores Technol 214:371–375. https://doi.org/10.1016/j.biortech.2016.04.113

Chang KL, Chen XM, Wang XQ, Han YJ, Potprommanee L, Liu JY, Liao YL, Ning XA, Sun SY, Huang Q (2017) Impact of surfactant type for ionic liquid pretreatment on enhancing delignification of rice straw. Biores Technol 227:388–392. https://doi.org/10.1016/j.biortech.2016.11.085

Chen WH, Chen YC, Lin JG (2014) Study of chemical pretreatment and enzymatic saccharification for producing fermentable sugars from rice straw. Bioprocess Biosyst Eng 37(7):1337–1344. https://doi.org/10.1007/s00449-013-1106-0

Chuetor S, Luque R, Barron C, Solhy A, Rouau X, Barakat A (2015) Innovative combined dry fractionation technologies for rice straw valorization to biofuels. Green Chem 17(2):926–936. https://doi.org/10.1039/C4GC01718H

Das A, Paul T, Jana A, Halder SK, Ghosh K, Maity C, Mohapatra PKD, Pati BR, Mondal KC (2013) Bioconversion of rice straw to sugar using multizyme complex of fungal origin and subsequent production of bioethanol by mixed fermentation of *Saccharomyces cerevisiae* MTCC 173 and *Zymomonas mobilis* MTCC 2428. Ind Crops Prod 46:217–225. https://doi.org/10.1016/j.indcrop.2013.02.003

De Bhowmick G, Sarmah AK, Sen R (2017) Lignocellulosic biorefinery as a model for sustainable development of biofuels and value added products. Biores Technol

de Lima Brossi MJ, Jiménez DJ, Cortes-Tolalpa L, van Elsas JD (2016) Soil-derived microbial consortia enriched with different plant biomass reveal distinct players acting in lignocellulose degradation. Microbial Ecol 71(3):616–627. https://doi.org/10.1007/s00248-015-0683-7

de Souza AP, Leite DCC, Pattathil S, Hahn MG, Buckeridge MS (2013) Composition and structure of sugarcane cell wall polysaccharides: implications for second-generation bioethanol production. BioEnergy Res 6(2):564–579

Devarapalli M, Atiyeh HK (2015) A review of conversion processes for bioethanol production with a focus on syngas fermentation. Biofuel Res J 2(3):268–280

Dhiman SS, Haw JR, Kalyani D, Kalia VC, Kang YC, Lee JK (2015) Simultaneous pretreatment and saccharification: green technology for enhanced sugar yields from biomass using a fungal consortium. Biores Technol 179:50–57. https://doi.org/10.1016/j.biortech.2014.11.059

Ding MZ, Song H, Wang EX, Liu Y, Yuan YJ (2016) Design and Construction of Synthetic Microbial Consortia in China. Synthetic and Systems Biotechnology 1(4):230–235. https://doi.org/10.1016/j.synbio.2016.08.004

Gong G, Liu D, Huang Y (2010) Microwave-assisted organic acid pretreatment for enzymatic hydrolysis of rice straw. Biosyst Eng 107(2):67–73. https://doi.org/10.1016/j.biosystemseng.2010.05.012

Gupta A, Verma JP (2015) Sustainable bio-ethanol production from agro-residues: a review. Renew Sustain Energy Rev. https://doi.org/10.1016/j.rser.2014.08.032

Gusakov AV, Sinitsyn AP, Klyosov AA (1985) Kinetics of the enzymatic hydrolysis of cellulose: 1. A mathematical model for a batch reactor process. Enzyme Microbial Technol 7(7):346–352. https://doi.org/10.1016/0141-0229(85)90114-0

Ibrahim MM, El-Zawawy WK, Abdel-Fattah YR, Soliman NA, Agblevor FA (2011) Comparison of alkaline pulping with steam explosion for glucose production from rice straw. Carbohyd Polym 83(2):720–726. https://doi.org/10.1016/j.carbpol.2010.08.046

Imman S, Arnthong J, Burapatana V, Champreda V, Laosiripojana N (2015) Fractionation of rice straw by a single-step solvothermal process: effects of solvents, acid promoters, and microwave treatment. Renew Energy 83:663–673

Jia X, Liu C, Song H, Ding M, Du J, Ma Q, Yuan Y (2016) Design, analysis and application of synthetic microbial consortia. Synth Syst Biotechnol 1(2):109–117. https://doi.org/10.1016/j.synbio.2016.02.001

Jin X, Ning Y (2013) Laccase production optimization by response surface methodology with *Aspergillus fumigatus* AF1 in unique inexpensive medium and decolorization of different dyes with the crude enzyme or fungal pellets. J Hazard Mater 262:870–877. https://doi.org/10.1016/j.jhazmat.2013.09.024

Jönsson LJ, Martín C (2016) Pretreatment of lignocellulose: formation of inhibitory by-products and strategies for minimizing their effects. Biores Technol 199:103–112

Kadam KL, McMillan JD (2003) Availability of corn stover as a sustainable feedstock for bioethanol production. Biores Technol. https://doi.org/10.1016/S0960-8524(02)00269-9

Kapoor M, Soam S, Agrawal R, Gupta RP, Tuli DK, Kumar R (2017) Pilot scale dilute acid pretreatment of rice straw and fermentable sugar recovery at high solid loadings. Biores Technol 224:688–693. https://doi.org/10.1016/j.biortech.2016.11.032

Karimi K, Emtiazi G, Taherzadeh MJ (2006) Ethanol production from dilute-acid pretreated rice straw by simultaneous saccharification and fermentation with *Mucor indicus*, *Rhizopus oryzae*, and *Saccharomyces cerevisiae*. Enzyme Microbial Technol 40(1):138–144. https://doi.org/10.1016/j.enzmictec.2005.10.046

Khaleghian H, Karimi K, Behzad T (2015) Ethanol production from rice straw by sodium carbonate pretreatment and *Mucor hiemalis* fermentation. Ind Crops Prod 76:1079–1085. https://doi.org/10.1016/j.indcrop.2015.08.008

Khaleghian H, Molaverdi M, Karimi K (2017) Silica removal from rice straw to improve its hydrolysis and ethanol production. Ind Eng Chem Res 56(35):9793–9798. https://doi.org/10.1021/acs.iecr.7b02830

Khoo HH (2015) Review of bio-conversion pathways of lignocellulose-to-ethanol: sustainability assessment based on land footprint projections. Renew Sustain Energy Rev. https://doi.org/10.1016/j.rser.2015.02.027

Kim JH, Block DE, Shoemaker SP, Mills DA (2010) Conversion of rice straw to bio-based chemicals: an integrated process using lactobacillus brevis. Appl Microbiol Biotechnol 86(5):1375–1385

Klein M, Griess O, Pulidindi IN, Perkas N, Gedanken A (2016) Bioethanol production from *Ficus religiosa* leaves using microwave irradiation. J Environ Manag 177:20–25

Ko JK, Bak JS, Jung MW, Lee HJ, Choi IG, Kim TH, Kim KH (2009) Ethanol Production from rice straw using optimized aqueous-ammonia soaking pretreatment and simultaneous saccharification and fermentation processes. Biores Technol 100(19):4374–4380. https://doi.org/10.1016/j.biortech.2009.04.026

Kogo T, Yoshida Y, Koganei K, Matsumoto H, Watanabe T, Ogihara J, Kasumi T (2017) Production of rice straw hydrolysis enzymes by the fungi *Trichoderma reesei* and *Humicola insolens* using rice straw as a carbon source. Biores Technol 233:67–73

Kumar P, Barrett DM, Delwiche MJ, Stroeve P (2009) Methods for pretreatment of lignocellulosic biomass for efficient hydrolysis and biofuel production. Ind Eng Chem Res 48:3713–3729

Kumar MN, Ravikumar R, Thenmozhi S, Sankar MK (2017) Development of natural cellulase inhibitor mediated intensified biological pretreatment technology using *Pleurotus florida* for maximum recovery of cellulose from paddy straw under solid state condition. Biores Technol 244:353–361

Lan TQ, Wei D, Yang ST, Liu X (2013) Enhanced cellulase production by *Trichoderma viride* in a rotating fibrous bed bioreactor. Biores Technol 133:175–182. https://doi.org/10.1016/j.biortech.2013.01.088

Lau BBY, Luis ET, Hossain MM, Hart WES, Cencia-Lay B, Black JJ, To TQ, Aldous L (2015) Facile, room-temperature pre-treatment of rice husks with tetrabutylphosphonium hydroxide: enhanced enzymatic and acid hydrolysis yields. Biores Technol 197:252–259

Lee KM, Kalyani D, Tiwari MK, Kim TS, Dhiman SS, Lee JK, Kim IW (2012) Enhanced enzymatic hydrolysis of rice straw by removal of phenolic compounds using a novel laccase from Yeast *Yarrowia lipolytica*. Biores Technol 123:636–645. https://doi.org/10.1016/j.biortech.2012.07.066

Lee C, Zheng Y, VanderGheynst JS (2015) Effects of pretreatment conditions and post-pretreatment washing on ethanol production from dilute acid pretreated rice straw. Biosys Eng 137:36–42. https://doi.org/10.1016/j.biosystemseng.2015.07.001

Lin TH, Huang CF, Guo GL, Hwang WS, Huang SL (2012) Pilot-scale ethanol production from rice straw hydrolysates using xylose-fermenting *Pichia stipitis*. Biores Technol 116:314–319. https://doi.org/10.1016/j.biortech.2012.03.089

López-Linares JC, Ballesteros I, Tourán J, Cara C, Castro E, Ballesteros M, Romero I (2015) Optimization of uncatalyzed steam explosion pretreatment of rapeseed straw for biofuel production. Biores Technol 190:97–105

Ludueña L, Fasce D, Alvarez VA, Stefani PM (2011) Nanocellulose from rice husk following alkaline treatment to remove silica. BioResources 6(2):1440–1453

Luo J, Fang Z, Smith RL (2014) Ultrasound-enhanced conversion of biomass to biofuels. Prog Energy Combust Sci. https://doi.org/10.1016/j.pecs.2013.11.001

Margot J, Bennati-Granier C, Maillard J, Blánquez P, Barry DA, Holliger C (2013) Bacterial versus fungal laccase: potential for micropollutant degradation. AMB Express 3:1–30. https://doi.org/10.1186/2191-0855-3-63

Mishima D, Kuniki M, Sei K, Soda S, Ike M, Fujita M (2008) Ethanol production from candidate energy crops: water hyacinth (*Eichhornia crassipes*) and water lettuce (*Pistia stratiotes* L.). Biores Technol 99:2495–2500. https://doi.org/10.1016/j.biortech.2007.04.056

Mofijur M, Masjuki HH, Kalam MA, Ashrafur Rahman SM, Mahmudul HM (2015) Energy Scenario and biofuel policies and targets in ASEAN countries. Renew Sustain Energy Rev. https://doi.org/10.1016/j.rser.2015.02.020

Mood SH, Golfeshan AH, Tabatabaei M, Jouzani GS, Najafi GH, Gholami M, Ardjmand M (2013) Lignocellulosic biomass to bioethanol, a comprehensive review with a focus on pretreatment. Renew Sustain Energy Rev 27:77–93. https://doi.org/10.1016/j.rser.2013.06.033

Moradi F, Amiri H, Soleimanian-Zad S, Ehsani MR, Karimi K (2013) Improvement of acetone, butanol and ethanol production from rice straw by acid and alkaline pretreatments. Fuel 112:8–13. https://doi.org/10.1016/j.fuel.2013.05.011

Mustafa AM, Poulsen TG, Xia Y, Sheng K (2017) Combinations of fungal and milling pretreatments for enhancing rice straw biogas production during solid-state anaerobic digestion. Biores Technol 224:174–182. https://doi.org/10.1016/j.biortech.2016.11.028

Nakanishi A, Bae JG, Fukai K, Tokumoto N, Kuroda K, Ogawa J, Nakatani M, Shimizu S, Ueda M (2012) Effect of pretreatment of hydrothermally processed rice straw with laccase-displaying yeast on ethanol fermentation. Appl Microbiol Biotechnol 94(4):939–948. https://doi.org/10.1007/s00253-012-3876-8

Nguyen TAD, Kim KR, Han SJ, Cho HY, Kim JW, Park SM, Park JC, Sim SJ (2010) Pretreatment of rice straw with ammonia and ionic liquid for lignocellulose conversion to fermentable sugars. Biores Technol 101:7432–7438. https://doi.org/10.1016/j.biortech.2010.04.053

Nguyen TY, Cai CM, Kumar R, Wyman CE (2017) Overcoming factors limiting high-solids fermentation of lignocellulosic biomass to ethanol. In: Proceedings of the National Academy of Sciences, 201704652

Okamoto K, Nitta Y, Maekawa N, Yanase H (2011) Direct ethanol production from starch, wheat bran and rice straw by the white rot fungus *Trametes hirsuta*. Enzyme Microbial Technol 48(3):273–277. https://doi.org/10.1016/j.enzmictec.2010.12.001

Parenti A, Muguerza E, Iroz AR, Omarini A, Conde E, Alfaro M, Castanera R, Santoyo F, Ramírez L, Pisabarro AG (2013) Induction of laccase activity in the white rot fungus *Pleurotus ostreatus* using water polluted with wheat straw extracts. Biores Technol 133:142–149. https://doi.org/10.1016/j.biortech.2013.01.072

Park JY, Shiroma R, Al-Haq MI, Zhang Y, Ike M, Arai-Sanoh Y, Ida A, Kondo M, Tokuyasu K (2010) A novel lime pretreatment for subsequent bioethanol production from rice straw—calcium capturing by carbonation (CaCCO) process. Biores Technol 101(17):6805–6811. https://doi.org/10.1016/j.biortech.2010.03.098

Phitsuwan P, Sakka K, Ratanakhanokchai K (2013) Improvement of lignocellulosic biomass in planta: a review of feedstocks, biomass recalcitrance, and strategic manipulation of ideal plants designed for ethanol production and processability. Biomass Bioenerg 58:390–405

Phitsuwan P, Permsriburasuk C, Baramee S, Teeravivattanakit T, Ratanakhanokchai K (2017) Structural analysis of alkaline pretreated rice straw for ethanol production. Int J Polym Sci. https://doi.org/10.1155/2017/4876969

Poornejad N, Karimi K, Behzad T (2013) Improvement of saccharification and ethanol production from rice straw by NMMO and [BMIM][OAc] pretreatments. Ind Crops Prod 41(1):408–413. https://doi.org/10.1016/j.indcrop.2012.04.059

Postemsky PD, Bidegain MA, González-Matute R, Figlas ND, Cubitto MA (2017) Pilot-scale bioconversion of rice and sunflower agro-residues into medicinal mushrooms and laccase enzymes through solid-state fermentation with *Ganoderma lucidum*. Biores Technol 231:85–93. https://doi.org/10.1016/j.biortech.2017.01.064

Qureshi N, Singh V, Liu S, Ezeji TC, Saha BC, Cotta MA (2014) Process integration for simultaneous saccharification, fermentation, and recovery (SSFR): production of butanol from corn stover using *Clostridium beijerinckii* P260. Biores Technol 154:222–228. https://doi.org/10.1016/j.biortech.2013.11.080

Rastogi S, Soni Raman, Kaur J, Soni SK (2016) Unravelling the capability of *Pyrenophora phaeocomes* S-1 for the production of ligno-hemicellulolytic enzyme cocktail and simultaneous bio-delignification of rice straw for enhanced enzymatic saccharification. Biores Technol 222:458–469. https://doi.org/10.1016/j.biortech.2016.10.012

Reactor, Tube, and Biochemical Engineering (2001) In design of a batch tube reactor for biomass hydrolysis 91(1):377–386

Saha BC, Yoshida T, Cotta MA, Sonomoto K (2013) Hydrothermal pretreatment and enzymatic saccharification of corn stover for efficient ethanol production. Ind Crops Prod 44:367–372. https://doi.org/10.1016/j.indcrop.2012.11.025

Saha BC, Qureshi N, Kennedy GJ, Cotta MA (2016) Biological pretreatment of corn stover with white-rot fungus for improved enzymatic hydrolysis. Int Biodeterior Biodegradation 109:29–35

Salvachúa D, Prieto A, López-Abelairas M, Lu-Chau T, Martínez ÁT, Martínez MJ (2011) Fungal pretreatment: an alternative in second-generation ethanol from wheat straw. Biores Technol 102:7500–7506. https://doi.org/10.1016/j.biortech.2011.05.027

Sarkar N, Ghosh SK, Bannerjee S, Aikat K (2012) Bioethanol production from agricultural wastes: an overview. Renew Energy. https://doi.org/10.1016/j.renene.2011.06.045

Shen F, Li H, Wu X, Wang Y, Zhang Q (2018) Effect of organic loading rate on anaerobic co-digestion of rice straw and pig manure with or without biological pretreatment. Biores Technol 250:155–162

Shinozaki Y, Kitamoto HK (2011) Ethanol production from ensiled rice straw and whole-crop silage by the simultaneous enzymatic saccharification and fermentation process. J Biosci Bioeng 111(3):320–325. https://doi.org/10.1016/j.jbiosc.2010.11.003

Silva JPA, Carneiro LM, Roberto IC (2013) Treatment of rice straw hemicellulosic hydrolysates with advanced oxidative processes: a new and promising detoxification method to improve the bioconversion process. Biotechnol Biofuels 6(1). https://doi.org/10.1186/1754-6834-6-23

Sindhu R, Kuttiraja M, Binod P, Janu KU, Sukumaran RK, Pandey A (2011) Dilute acid pretreatment and enzymatic saccharification of sugarcane tops for bioethanol production. Biores Technol 102:10915–10921. https://doi.org/10.1016/j.biortech.2011.09.066

Sindhu R, Binod P, Pandey A (2016) Biological pretreatment of lignocellulosic biomass–an overview. Biores Technol 199:76–82

Sindhu R, Binod P, Mathew AK, Abraham A, Gnansounou E, Ummalyma SB, Thomas L, Pandey A (2017) Development of a novel ultrasound-assisted alkali pretreatment strategy for the production of bioethanol and xylanases from chili post harvest residue. Biores Technol

Singh A, Tuteja S, Singh N, Bishnoi NR (2011) Enhanced saccharification of rice straw and hull by microwave-alkali pretreatment and lignocellulolytic enzyme production. Biores Technol 102(2):1773–1782. https://doi.org/10.1016/j.biortech.2010.08.113

Singh R, Tiwari S, Srivastava M, Shukla A (2014) Microwave assisted alkali pretreatment of rice straw for enhancing enzymatic digestibility. J Energy 2014:1–7. https://doi.org/10.1155/2014/483813

Singh J, Suhag M, Dhaka A (2015) Augmented digestion of lignocellulose by steam explosion, acid and alkaline pretreatment methods: a review. Carbohyd Polym 117:624–631

Singh R, Srivastava M, Shukla A (2016) Environmental sustainability of bioethanol production from rice straw in india: a review. Renew Sustain Energy Rev. https://doi.org/10.1016/j.rser.2015.10.005

Soam S, Kapoor M, Kumar R, Borjesson P, Gupta RP, Tuli DK (2016) Global warming potential and energy analysis of second generation ethanol production from rice straw in India. Appl Energy 184:353–364. https://doi.org/10.1016/j.apenergy.2016.10.034

Sun S, Sun S, Cao X, Sun R (2016) The role of pretreatment in improving the enzymatic hydrolysis of lignocellulosic materials. Biores Technol 199:49–58

Swain MR, Krishnan C (2015) Improved conversion of rice straw to ethanol and xylitol by combination of moderate temperature ammonia pretreatment and sequential fermentation using *Candida tropicalis*. Ind Crops Prod 77:1039–1046. https://doi.org/10.1016/j.indcrop.2015.10.013

Toquero C, Bolado S (2014) Effect of four pretreatments on enzymatic hydrolysis and ethanol fermentation of wheat straw. Influence of inhibitors and washing. Biores Technol 157:68–76. https://doi.org/10.1016/j.biortech.2014.01.090

Umagiliyage AL, Choudhary R, Liang Y, Haddock J, Watson DG (2015) Laboratory scale optimization of alkali pretreatment for improving enzymatic hydrolysis of sweet sorghum bagasse. Ind Crops Prod 74:977–986

Union, Europäische (2009) Directive 2009/28/EC of the European Parliament and of the Council of 23 April 2009 on the promotion of the use of energy from renewable sources and amending and subsequently repealing directives 2001/77/EC and 2003/30/EC. Official J European Union 5:2009

Valdez-Vazquez I, Pérez-Rangel M, Tapia A, Buitrón G, Molina C, Hernández G, Amaya-Delgado L (2015) Hydrogen and butanol production from native wheat straw by synthetic microbial consortia integrated by species of Enterococcus and Clostridium. Fuel 159:214–222

Vincent M, Pometto AL, Van Leeuwen JH (2014) Ethanol production via simultaneous saccharification and fermentation of sodium hydroxide treated corn stover using *Phanerochaete chrysosporium* and *Gloeophyllum trabeum*. Biores Technol 158:1–6. https://doi.org/10.1016/j.biortech.2014.01.083

Wan C, Li Y (2010) Microbial pretreatment of corn stover with *Ceriporiopsis subvermispora* for enzymatic hydrolysis and ethanol production. Biores Technol 101(16):6398–6403

Wi SG, Choi IS, Kim KH, Kim HM, Bae HJ (2013) Bioethanol production from rice straw by popping pretreatment. Biotechnol Biofuels 6(1). https://doi.org/10.1186/1754-6834-6-166

Wu J, Zhang X, Wan J, Ma F, Tang Y, Zhang X (2011) Production of fiberboard using corn stalk pretreated with white-rot fungus *Trametes hirsute* by hot pressing without adhesive. Biores Technol 102(24):11258–11261

Wu H, Dai X, Zhou SL, Gan YY, Xiong ZY, Qin YH, Ma J, Yang L, Wu ZK, Wang TL (2017) Ultrasound-assisted alkaline pretreatment for enhancing the enzymatic hydrolysis of rice straw by using the heat energy dissipated from ultrasonication. Biores Technol 241:70–74

Xia A, Cheng J, Song W, Yu C, Zhou J, Cen K (2013) Enhancing enzymatic saccharification of water hyacinth through microwave heating with dilute acid pretreatment for biomass energy utilization. Energy 61:158–166

Xing MN, Zhang XZ, Huang H (2012) Application of metagenomic techniques in mining enzymes from microbial communities for biofuel synthesis. Biotechnol Adv. https://doi.org/10.1016/j.biotechadv.2012.01.021

Zahed O, Jouzani GS, Abbasalizadeh S, Khodaiyan F, Tabatabaei M (2016) Continuous co-production of ethanol and xylitol from rice straw hydrolysate in a membrane bioreactor. Folia Microbiol 61(3):179–189

Zhao J, Xia L (2010) Ethanol production from corn stover hemicellulosic hydrolysate using immobilized recombinant yeast cells. Biochem Eng J 49:28–32. https://doi.org/10.1016/j.bej.2009.11.007

Zheng YG, Chen XL, Wang Z (2005) Microbial biomass production from rice straw hydrolysate in airlift bioreactors. J Biotechnol 118(4):413–420. https://doi.org/10.1016/j.jbiotec.2005.04.022

Zhu S, Huang W, Huang W, Wang K, Chen Q, Wu Y (2015) Pretreatment of rice straw for ethanol production by a two-step process using dilute sulfuric acid and sulfomethylation reagent. Appl Energy 154:190–196. https://doi.org/10.1016/j.apenergy.2015.05.008

Zuroff TR, Curtis WR (2012) Developing symbiotic consortia for lignocellulosic biofuel production. Appl Microbiol Biotechnol 93(4):1423–1435

https://www.afdc.energy.gov/data/10331

http://www.energetica-india.net/articles/renewable-energy (2009)

http://nuicone.org/site/common/proceedings/Chemical/oral/CH_21.pdf

http://biofuel.org.uk/types-of-biofuels.html

Integrated Lignocellulosic Biorefinery for Sustainable Bio-Based Economy

Jitendra Kumar Saini, Rishi Gupta, Hemansi, Amit Verma, Priyanka Gaur, Ritu Saini, Rishikesh Shukla and Ramesh Chander Kuhad

Abstract Upsurge of interests in biomass-based economy has opened up many new challenges related to knowledge, technology, economics and society. The complexity of lignocellulosic biomass is comparable to petroleum, and hence, the concept of biorefinery has emerged. A better understanding of the complexity of the lignocellulosic biomass has helped in exploitation of each of its constituent at the fullest for production of a wide variety of products in comparison to the production of single products previously. The production of multiple products such as biofuels and other valuable bio-based materials from lignocellulosic feedstock requires integration of various processes in biorefinery operations in analogy with petroleum-based refineries. Therefore, it is important to understand key issues of lignocellulose biorefining. This chapter deals with the concept and practice of integrated lignocellulosic biorefinery for sustainable development, different products and the sustainability aspects.

J. K. Saini (✉) · Hemansi · P. Gaur · R. Shukla
Department of Microbiology, School of Interdisciplinary & Applied Life Sciences, Central University of Haryana, Mahendergarh 123031, Haryana, India
e-mail: me_jk@rediffmail.com; jitendrasaini@cuh.ac.in

R. Gupta
Department of Biotechnology, School of Interdisciplinary & Applied Life Sciences, Central University of Haryana, Mahendergarh 123031, Haryana, India

A. Verma
Department of Biochemistry, College of Basic Sciences and Humanities, S D Agricultural University, SK Nagar, Gujarat, India

R. Saini
Department of Bioenergy, DBT-IOC Centre for Advanced Bioenergy Research, Faridabad 121007, India

R. C. Kuhad
Department of Microbiology, University of Delhi South Campus, New Delhi 110021, Delhi, India

R. C. Kuhad
Central University of Haryana, Mahendergarh 123031, Haryana, India

Present Address
R. Gupta
SVI Research, SVI Analytica, Gurugram 122016, Haryana, India

N. Srivastava et al. (eds.), *Sustainable Approaches for Biofuels Production Technologies*, Biofuel and Biorefinery Technologies 7,
https://doi.org/10.1007/978-3-319-94797-6_2

In the end, current status and future prospects of lignocellulosic biomass-based biorefinery are discussed.

Keywords Biorefinery · Biofuel · Lignocellulose · Sustainability · Biomass

1 Introduction

Presently, we are almost fully dependent upon exhaustible supplies of petroleum, coal and other fossil fuels for meeting the day to day demands of transportation fuels, energy, commodity chemicals and various other products. Due to rapid and ever-increasing consumption of energy and commodity chemicals, search for sustainable supplies of carbon-neutral feedstock and/or resources as well as production processes is inevitable (Maity 2015). In biorefinery, production of biofuels, renewable energy/power, and various chemicals from plant biomass resources is carried out based upon the integration of various production methods as well as the machinery (Luo et al. 2010). The concept of biorefinery can be considered equivalent to that of petrochemical refinery in a way that the biomass feedstock is processed to obtain a multitude of products, such as biofuels, chemicals, biomaterials, biomolecules, etc. (Moncada et al. 2015).

A variety of physico-chemical, biochemical/microbial processes are used in biorefinery operations for production of various products to be used in transportation, pharmaceutical/health, food and other sectors, with zero or minimal waste generation. Complete utilisation of biomass for large volume production of low commercial value product simultaneously with one or more high-value products will enhance the competitiveness of biorefinery operation (https://www.aber.ac.uk/en/media/departmental/ibers/pdf/innovations/07/07ch8.pdf accessed on 20th Dec 2017).

The production of transportation fuels and chemicals from each and every fraction and/or process wastes during biomass processing is thus an essential requirement of an integrated biorefinery and has the potential to reduce dependence upon non-renewable fossil-based resources and decline global warming as well. Sustainability of the biorefineries requires that the production of multi-products in total should be economic, energy efficient, environmentally safer and carbon-neutral (or preferably carbon negative). Biorefinery is expected to produce a variety of products both in terms of their chemical/biochemical properties and their economic or commercial value. To be economic, it is desirable that the volumes of the low-value products such as bioethanol, biodiesel, etc. should be high; however, the low-cost chemicals and biomaterials obtained in even low volumes can give similar or higher economic advantage (Moncada et al. 2015). Although integrated biorefinery concept is gaining worldwide publicity and acceptance and has enormous potential as far as the sustainable and green manufacture of bio-based products is concerned, this area is still in its infancy and comparatively fewer reports have focused on all of its aspects. Moreover, due to newer scientific as well as technological advances in this ever-growing area (Kamm and Kamm 2004; Maity 2015) the concepts of biorefinery are evolving and being redefined very fast. Therefore, there is a need to understand various concepts of biorefinery in more detail that will be helpful in developing a techno-socio-economically viable biorefinery system. This chapter deals with the concept and practice of integrated lignocellulosic biorefinery for sustainable development, different products of integrated lignocellulosic biorefinery and process technology in lignocellulosic biorefinery with important considerations. In the end, a short survey of the biorefinery industry followed by challenges, opportunities and future prospects is provided.

2 Biorefinery Concept

The perception of biorefinery developed towards the end of the twentieth century to address the challenges of declining petroleum-based transportation fuels and outlook for cheaper and environmentally friendly synthesis of valuable materials and products from renewable biomass (Kamm and Kamm 1997; Kamm et al. 2007). The concept of biorefinery is depicted in Fig. 1.

The concept of biorefinery is similar to petrochemical/refinery industry in a manner that both types of refineries are involved in production of a wide array of chemicals as well as fuels (Maity 2015). Globally, revenue generation by bio-based products

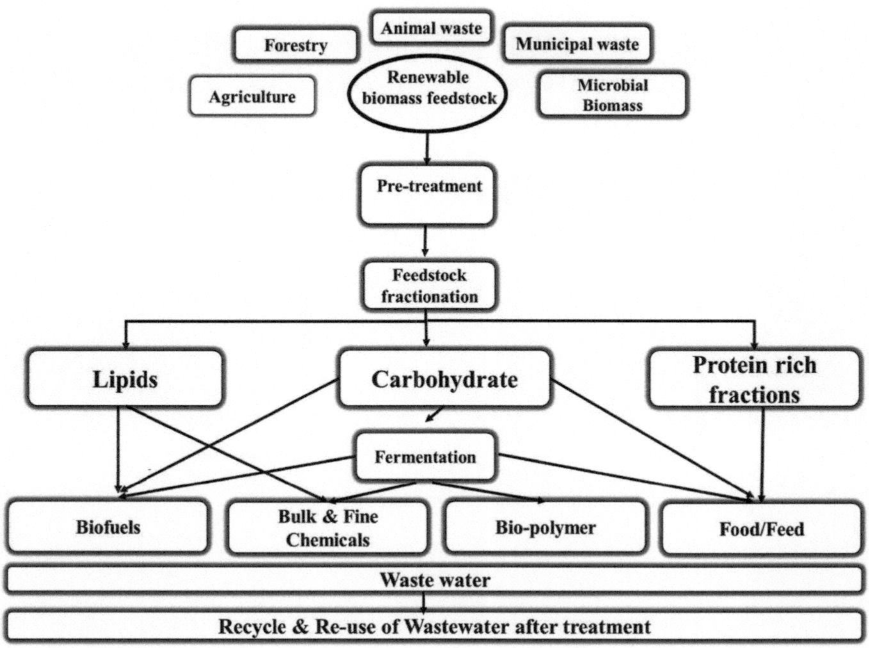

Fig. 1 The concept of biorefinery

for chemical industry is estimated at USD 10–15 billion. Many papers, reviews and reports have addressed the biomass potential for chemical and polymer production in much detail (de Jong et al. 2012). Some comparable aspects of biorefinery petroleum refinery are engineering aspects, including feedstock fractionation, multiple products (both platform and end use), process integration and flexibility as depicted in Fig. 2. Two major differences are the raw material and complexity in application of technologies. Some of the products produced in the biorefinery cannot be obtained after refining petroleum, e.g. some food products (Moncada et al. 2016).

Need for biorefinery in current scenario is threefold: (1) the depletion of petroleum resources, (2) concerns about global climate change and (3) energy security issues. Additionally, there are other reasons for the need of biorefineries, such as avoiding over-dependence on petroleum-based products; need of economic products and chemicals; strengthening bio-based circular economy; protection of natural environment and ecosystem; and increasing employability in rural regions and stimulate the sustainable development of regional areas (Langeveld et al. 2012; McCormick and Kautto 2013). There are a number of merits of biorefinery-based production of bioenergy, biofuel, biochemicals and materials as shown in Fig. 3.

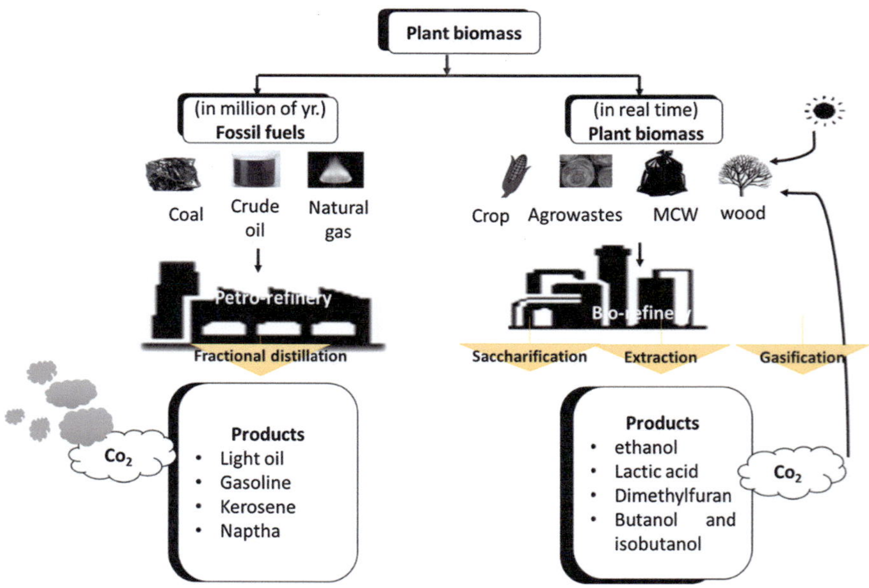

Fig. 2 Analogy between petroleum-based refinery and biorefinery

3 Type of Biorefineries

In general, biorefineries are categorised into three different classes depending upon the utilisation of the substrate: first-generation (starch and sugar feedstock) biorefineries, second-generation (plant biomass feedstock) biorefineries and third-generation (algal feedstock) biorefineries (Hossain et al. 2016). These are discussed below in more detail, and the examples of each type of these biorefineries and respective products are listed in Table 1.

3.1 First-Generation Biorefineries

Such biorefineries are utilising sugary, oil-based and starch-containing substrates as feedstocks to produce fuels and commodity chemicals. Currently, almost all biofuels including bioethanol and biodiesel, and biochemicals are produced by first-generation biorefinery. The main drawback of these biorefineries is competition of raw substrate with food demand and deterioration of soil.

3.2 Second-Generation Biorefineries

In these, the feedstock is lignocellulose-containing plant or forestry-based biomass such as agricultural residue, forestry and urban waste. The complexity of chemical composition of lignocellulosic biomass is the main reason for its usefulness in

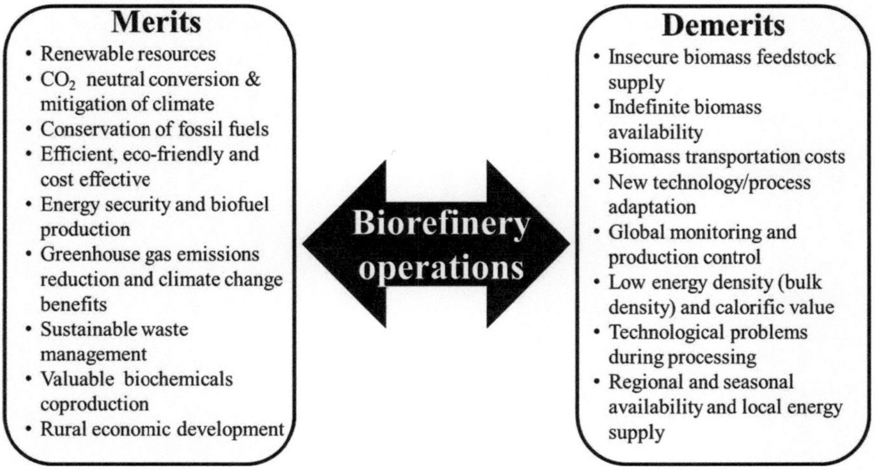

Fig. 3 Merits and demerits of lignocellulosic biorefineries

Table 1 Examples of types of biorefineries and their products

Type of biorefinery	Substrate type	Process	Products	References
First-generation biorefinery	Corn (sugar-based)	Wet milling followed by enzymatic saccharification and fermentation	**Mainstream products**: starch, glucose, hemicellulose, corn meal, gluten and corn protein **Derived products**: **Hemicellulose**—glucose, xylose and arabinose **Xylose**—xylitol, starch acetate-corn fibre foam **Corn gluten meal**—feed, food and pharma-based compounds **Gluten**—corn gluten hydrolysates, plastics and composites **Corn proteins**—films/coatings, food and beverages additives, and corn steep liquor	(Amartey and Jeffries 1994; Shukla and Cheryan 2001; De Azeredo et al. 2006; Gennadios 2002; Ganjyal et al. 2004; Kim et al. 2004; Jerez et al. 2005; Aithani and Mohanty 2006; Kılıç Apar and Ozbek 2007; Agarwal et al. 2008; Samarasinghe et al. 2008)
	Soybean (lipid-based)	High-temperature and pressure-based alcohol technology Enzymatic catalytic technology	**Mainstream products**: glycerol, soybean meal and soybean oil **Derived products**: **Glycerol**—polyhydroxyalkanoates, various organic acids, industrial solvents, platform chemicals and many sugar alcohols **Soybean meal**—animal feeds, lipopeptides, PGA and isoflavones **Soybean oil**—lipase, lecithin	(Szuhaj 1989; Vancauwenberge et al. 1990; Berk 1992; Eggink et al. 1994; Barbirato et al. 1997; Bormann and Roth 1999; Demirbas 2005; Doleyres et al. 2005; Behzadi and Farid 2007; Al-Zuhair 2007; Anand and Saxena 2012)
Second-generation biorefinery	Sugarcane bagasse (lignocellulose-based)	Pretreatment, hydrolysis and fermentation	**Main products**: sugar, bagasse and ethanol **Derived products**: **Sugars**: ethanol, acids, poly (3-hydroxybutyric acid) and other chemicals **Bagasse**: cellulose, xylan, lignin, sugars, ethanol, chemicals, compost, electricity and steam	(Brumbley et al. 2007; Nass et al. 2007; Nel 2010)
Third-generation biorefinery	Microalgae (algal-based)	Extraction hydrolysis and fermentation	**Main products**: pulp, lipids, cellulose, agar, kappa-carrageenan, etc. **Derived products**: **Lipids**: biodiesel **Cellulose**: sugars, alcohols, acids, chemicals **Agar:** agarose **Pulp:** animal feed	(Lakaniemi et al. 2013)

manufacturing of various chemicals as well as fuels (Maity 2015). The biomass can be processed through thermochemical as well as biological routes. The abundance, diversity and non-competence with food crops of lignocellulosic biomass make it superior to first-generation biorefineries.

3.3 Third-Generation Biorefineries

Microalgae, as third-generation substrate, have vast potential for the sustainable production of commodity products. Such biorefineries can provide cleaner energy (biodiesel and bioethanol), value-added products including cosmetics, therapeutics, animal feed and food, and technical solution to waste management concerns. The major advantage of using microalgae as a substrate is their ability to grow very fast within a shorter span of time.

4 Lignocellulosic Biorefineries

Due to the issues of economic sustainability and environmental concerns, the global research interest for production of various chemicals, fuels, energy and other materials has been shifted towards renewable sources as substitutes to petroleum-derived products. Lignocellulose biomass, the most abundantly available organic carbon source, can be a sustainable alternative to petroleum-dependent fuels and petrochemicals and will surely emerge as an important source of biomass and be widely available at moderate costs showing less competition with food and feed production.

4.1 Feedstock and Products

Lignocellulosic biorefinery under different above said categories are mentioned in Table 2. Lignocellulosic biomass is the most abundant biomass with vast potential for production of a wide range of bio-products and biofuels (Amidon and Liu 2009; Liu et al. 2012; Menon and Rao 2012; Maity 2015). The annual production of lignocellulosic biomass has been reported to be approximately 150–170 $\times$ 10^9 tonnes. However, despite its abundance and low cost, the conversion of lignocellulosic biomass to value-added products and their selective recovery remain a bottleneck due to the lack of economic viability, and this has become the active area for extensive research across the globe to address this concern worldwide (Cherubini and Ulgiati 2010; Sarma et al. 2017).

Lignocellulosic biomass broadly can be categorised into agriculture waste, forest and industrial waste, aquatic waste and municipal waste. Included among the first category are various crop wastes such as straws (wheat, rice), stalks (cotton,

Table 2 Biomass feedstock for lignocellulosic biorefineries

Agricultural residues	Woody biomass	Aquatic biomass	Municipal solid wastes
Wheat and rice straw, corn cobs, stalks of cotton plant and barley, empty fruit bunch from oil palm, corn cob, maize and sorghum stover, peanut shell, and bagasse	*Prosopis juliflora*, *Lantana camara*, pine needles, *Saccharum munda*, willow, eucalyptus, beech wood, birch wood, cedar wood, pinewood, Douglas fir and oak bark	*Gracilaria verrucosa*, giant kelp, *Kappaphycus* sp., *Sargassum* sp., diatoms, *Ulva lactuca*, *Chlorococcum* sp., *Porphyra* sp., *Palmaria* sp.	A variety of biomass and organic materials (majorly from kitchen waste), from which lignocellulosic solid waste can be used after segregation

mustard) and bagasse (cane, sweet sorghum) that often are burned to prepare the agricultural fields for sowing of next crop. Such feedstock does not compete with food and is widely available. The main challenge in their exploitation in a biorefinery is their transportation cost in view of their low density and unavailability at a single place (Kamm and Kamm 2004). Biomass in the second category also has a similar composition to agricultural crop wastes, except that these have comparatively lower cellulose and more lignin (Kim et al. 2006; Speight 2014) and the biomass is not affected much by seasonal variations and the main reasons for biomass variations are location and forest type. Algal (aquatic) lignocellulosic biomass also does not compete with the food and is benefitted by much higher production rate under cheaper conditions (Talebian-Kiakalaieh et al. 2014), utilising either open ponds or photobioreactors (Nakamura and Whited 2003; Maity 2015). Potential biomass feedstocks for lignocellulosic biorefinery under different above said categories are mentioned in Table 2.

Lignocellulose mainly contains three complex structural polymer entities including cellulosic, hemicellulosic and lignin fractions (Fig. 4), which can be utilised for synthesis of various useful chemicals and products using enzymatic/biochemical or chemical platform after their conversion to simpler sugars. Lignin is another polymer of high economic importance, which can be utilised for cogeneration; synthesis of phenolic components and other chemicals. (de Bhowmick et al. 2018). Different products obtained from various factions of lignocellulosic biomass are depicted in Fig. 5 and these products are categorised into five different types, i.e. biofuels, bioenergy, food products, biochemical and biomaterials in Table 3.

Interestingly, via lignocellulosic biorefinery-based approach, various platform chemicals can be formed by direct fermentation of sugars [ethanol (C2); propanol (C3); butanol (C4)], syngas transformation of propylene (C-3), dimerization of ethylene (produced from dehydration of ethanol) to butenes (C-4), etc. Moreover, approaches such as ABE fermentation can also be used to produce multiple products acetone, butanol and ethanol (BREW 2006; Bos and Sanders 2013; Yao and Tang 2013; Kajaste 2014). A list of examples of C1-C6 platform chemicals derived from lignocellulosic biomass through microbial fermentation is shown in Table 4.

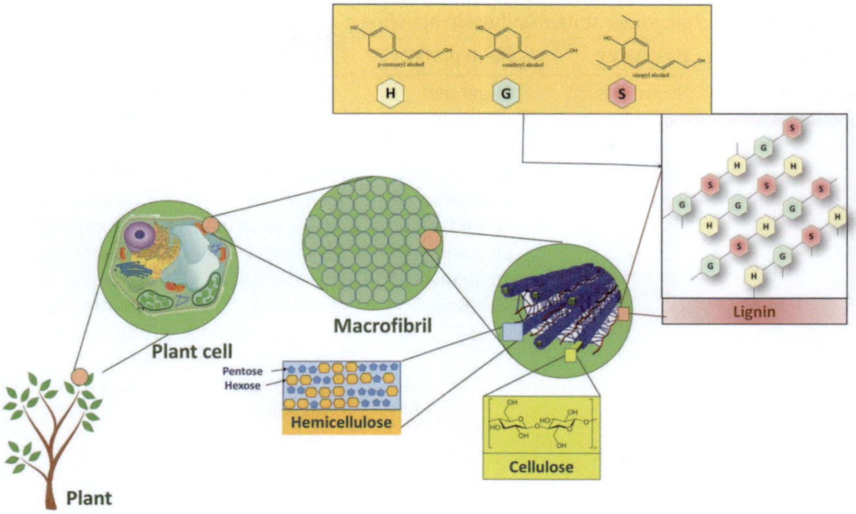

Fig. 4 Structural components of lignocellulosic biomass. H: p-coumaryl alcohol; G: coniferyl alcohol and S: sinapyl alcohol

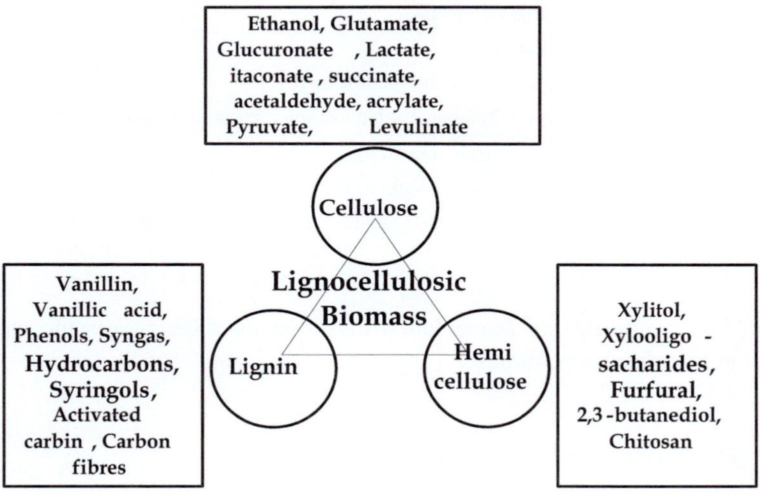

Fig. 5 Various commodity products derived from lignocellulosic components

4.2 Lignocellulose Conversion Processes

Depending upon the diversity of biomass and their compositional variability, a variety of conversion processes such as thermal and chemical (e.g. combustion, liquefaction, fast pyrolysis, etc.), chemical [viz. aqueous phase dehydration/hydrogenation

Table 3 Different categories of lignocellulosic biorefinery products (Moncada et al. 2016)

Biofuels	Bioenergy	Food products	Biochemicals	Biomaterials
Biodiesel, bioethanol and biomethane	Steam power, electricity steam, syngas, heat, charcoal and lignin	Sugar and substitutes, proteins, amino acids, gluten, protective colloids thickeners, emulsifiers and stabilisers	Simpler hexose and pentoses, and their degradation products such as 5-hydroxy methyl furfural, glycerol, agrochemicals, fertilisers, sorbitol, phenols, coloured compounds, solvents, omega-3 fatty acids and biosurfactants	Pulp and papers, PHB, activated carbon, bioplastics, bio-based epoxy, resin, cement, bioadhesives, bio-based polymers, bio-nanocomposites, etc.

Table 4 Typical chemicals from biomass produced via microbial fermentation

Carbon atoms	Platform chemicals produced via fermentation
C1	Methane (biogas), ethanol
C2	Acetate and lactate
C3	Glycerol, propanediol and fumaric acid
C4	Succinic acid, 1-butanol, 1,4-butanediol, aspartic acid and furfural
C5	Butanetriol, levulinate and citrate
C6	Lysine, glutamate and 5-hydroxymethylfurfural
Others	Amino acids, vitamins, biopolymers, industrial enzymes and antibiotics

(APD/H)], or biological (e.g. fermentation, digestion, microbial processing, etc.) can be used in biorefinery operations.

4.2.1 Thermochemical Processes

Such conversions include gasification process, fast pyrolysis and liquefaction processes (Balat 2008). The process of gasification generates various intermediates, for synthesis of various chemicals and biomaterials, as well as liquid biofuels, electricity and heat (Fahlén and Ahlgren 2009). In pyrolysis, the chemical reaction involves short interval high-temperature treatment of biological feedstock under anaerobic conditions. At low temperature, the process converts biomass to a liquid biocrude, which can subsequently be used to generate liquid biofuels (Balat 2008). Chemical reactions under liquefaction process are carried out under an environment of H_2O and CO/H_2 (Leduc et al. 2008; Wetterlund and Soderstrom 2010), for generation of bio-crudes and other important products (Palmqvist and Hahn-Hägerdahl 2000; Hamelinck et al. 2005).

4.2.2 Biochemical Processes

Such process consists of depolymerisation of structural polymers viz. cellulose and hemicellulose into monomeric sugars and further fermentations or enzymatic reactions for synthesis of useful products, such as bioethanol, biobutanol, ETBE, MTBE, acids, etc. The major drawback of thermochemical conversion of lignocellulosic biomass is high process cost (especially enzymes) and recalcitrance of biomass. However, biorefinery approach can compensate the cost by production of high-value products.

4.2.3 Chemical Processes

Several chemical processes including acid hydrolysis, can be used to intricate in lignocellulosic biorefinery. These can lie may be in pretreatment step or in downstream processing. Under controlled conditions, acid hydrolysis can convert lignocellulosic biomass into xylan, xylose and monomer sugars and could fractionate cellulose and lignin components (Palmqvist and Hahn-Hägerdal 2000). For chemical transformation of lignocellulosic biomass to syngas ($CO + H_2$), Fischer–Tropsch process can also be employed. Moreover, methyl alcohol production, hydro-formylation and methane synthesis can be carried out by using synthesis gases (Balat 2008).

4.3 Examples of Lignocellulosic Biorefineries

4.3.1 Bioethanol-Based Biorefinery

Cellulosic ethanol production involves various processes including its enzymatic after physico-chemical treatment followed by fermentation of hydrolysate and ethanol separation. Lignocellulose due to its structural complexity are very recalcitrant for its bioconversion and therefore, a prior physico-chemical processing or treatment step is carried out to remove biomass recalcitrance and make it amenable for enzymatic and microbial attack (Zheng et al. 2009). The pretreated substrate is then subjected to enzymatic depolymerisation, which is the most cost-intensive process due to high cost of enzymes. The hydrolysate thus obtained has both five-carbon and six-carbon sugars, which can be fermented to yield bioethanol. In nature, pentose fermenting microbes are very few and have relatively low yields of ethanol than C-6 fermenting microorganisms. In the final stage, ethanol thus produced after fermentation is harvested and concentrated by distilling the medium and/or by membrane separation. All these concerns led to a urge in developing bioethanol-based

Table 5 Comparative economics of lignocellulosic biorefinery and bioethanol plant (adapted from Luo et al. 2010)

S.No.	Item description	Economic analysis of biorefinery in comparison to bioethanol industry[a]
1	Capital investment	1.96
2	Variable operating costs	2.68
3	Fixed operating costs	1.96
4	NPV	28.39
5	IRR	3.53

[a]The numbers denote the magnitude of scale of costs and benefits in comparison to that of a bioethanol industry producing bioethanol as a single product

biorefinery so as to compensate the cost of bioethanol from the high-value additional products. Utilising ethanol yielding fraction such as hemicellulose and lignin to other value-added products could be a cost-effective approach to be considered. The solid residual unreacted products such as lignin, cellulose, hemicelluloses, enzymes and microorganisms are recovered after final ethanol recovery step and processed into other fuels (Mosier et al. 2005). Usually, solid residuals are dried to 10% mc and fired in a boiler or a gasifier to produce methane. A comparison of cost economic of bioethanol plant and bioethanol-based biorefinery is shown in Table 5.

4.3.2 Biomethane-Based Biorefinery

A number of crop residues including waste from maize, wheat, rye, etc. can be used as substrate for the production of biomethane. It is estimated that the annual maize and cereal crop waste has the potential to produce 2000–4500 MT of methane per hectare (Kumar et al. 2008). Similar to bioethanol production, biomethanation is also a multistep process. Methane fermentation from lignocellulosic biomass involves hydrolysis, acidogenesis, acetogenesis and methanation steps. The diversity of microorganisms required for each step varies from each other. Microorganisms through various phases, finally, hydrolyse the undissolved complex structural polymers of lignocellulose such as cellulose, proteins and fats into monomers. The monomeric sugars thus formed after hydrolysis are further exploited by other organisms to produce various C1–C5 molecules, alcohols, short chains of organic acids, hydrogen and CO_2 (Chandra et al. 2012). In the acetogenic phase, the organisms convert organic acids and alcohols into acetate. Finally, under the obligate anaerobes ferment these carbon sources (CO_2, formate, methanol and acetate, etc.) to methane in the methanogenesis. A variety of products formed in between can be recovered from the process and can be used to compensate the process cost. Moreover, the leftover biomass could be used as compost. Moreover, the process also offers a potential solution to the waste management.

4.3.3 Biohydrogen-Based Biorefinery

A number of lignocellulosic feedstocks including agriculture waste, stillage, industrial waste, fibre waste, kitchen waste, etc. are good feedstocks when seeking H_2 generation. Out of the available processes for biohydrogen production, the most cost-competitive process is usually the one involving only single stage. Usually, the lignocellulosic biomass is pretreated and hydrolysed followed by dark fermentation of the hydrolysate for hydrogen production. However, pretreatments may have drawbacks of generating undesirable by-product that could threat the fermentability of the hydrolyzates (Cheng et al. 2011; Quéméneur et al. 2012). Various interventions in this regards are undergoing and a major shift of research interest has been made recently. Cheng et al. (2011) have developed a novel process of biohydrogen production involving two stages comprising alternate light and dark phases using phototrophic microalgal strains. The process also underlines the requirement of integrating all the techniques to produce multiple products at a time.

5 Sustainability Aspects of Lignocellulosic Biorefineries

The initial thrust to the concept of 'sustainability' in relation to the environment was derived from 'The Brundtland report' of WCED as 'development that can meet the needs of the present generations without compromising abilities of future generations to meet their own demands' (Hofer and Bigorra 2008). In this context, through sustainable development we can preserve the quality of life for our coming generations. 'Sustainability' and 'sustainable development' are the two broader terms whose exact meaning and definitions are highly reliant on the milieu, specific goals and solicited use and may be considered multidimensional. In actual terms, 'sustainability' and 'sustainable development' can be considered to be associated with the balance of three important as well as interdependent aspects, i.e. econo-, enviro- and societal aspects, so that the well-being of our and our coming generations is preserved (Kemp and Martens 2007; Posada and Osseweijer 2016; Parada et al. 2017). The overall impacts of any biorefinery project can be realised in a real sense by wise combination of above-mentioned sustainability aspects after avoiding overlapping aspects and putting proper weightage to various indicators, subcategories and impacts categories (Santoyo-Castelazo and Azapagic 2014).

5.1 Economic Sustainability

This refers to the expenditures involved in each and every stage involved in biomass production, collection, processing, product formation, recovery, commercialization, etc. Therefore, economic sustainability takes into account the cost-competitiveness by combining the technical and economic aspects jointly. If the products are not cost-competitive, then most likely they will not have any market despite derived from renewable feedstocks (Posada and Osseweijer 2016). Economic indicators can be categorised into three classes associated with the cost, benefit and value of investment. The important economic indicators in the first category include capital cost, total savings, operating cost, production cost, transportation cost and actual sequestration cost, with production cost being the most critical indicator. The second category associated with benefit includes margins and profit related to operation among which latter one is the most important and most frequently used indicator. The third class of economic indicators includes indicators related to the investment value of a biorefinery and consists of return on investment, duration for payback, total economic value, NPV, stakeholder value and minimum selling price. Net present value is the most critical as well globally used indicator in this category followed by minimum selling price (Seider et al. 2010; Tan et al. 2016; Parada et al. 2017).

5.2 *Environmental Sustainability*

This aspect of sustainability of biorefineries helps in minimising the potential environmental hazards of biorefinery while producing the desired product in optimum quantity and without affecting the economic sustainability.

A systematic set of procedures for compiling and examining the inputs and outputs of materials and energy and the associated environmental impacts directly attributable to the functioning of a product or service system throughout its life cycle. Various phases of product formation right from derivation or synthesis from its source up to its final use/consumption constitutes its life cycle and evaluation of the product's life cycle for its impact on the environment are known as the life cycle assessment or LCA (ISO 2006). Additional efficient methodologies for environmental assessment may be based on evaluation of minimum impact, (Stefanis et al. 1995), minimum waste generation (Young and Cabezas 1999), risk to environment (Shonnard and Hiew 2000), thermodynamic analysis method (Bakshi 2002) and atmospheric hazards index (Gunasekara and Edwards 2003). Strategy for impact assessment may involve application of tools such as ReCiPe, CML, etc. based upon various factors related to several aspects of environment. However, modelling of some categories on basis of one geographic region may be inadequate for other geographical locations (Institute for Environment and Sustainability 2010). CO_2 emission has direct impact on environmental sustainability of biorefinery and has been mainly attributed to the production, utility generation and transportation in biorefinery applications (Parada et al. 2017). Other GHG emissions are also equally relevant in all stages of any biorefinery, including agricultural practices also (Fan et al. 2013).

5.3 Social Sustainability

This aspect determines the usefulness and implications of any biorefinery product, process or service. This dimension of sustainability has not been considered much frequently in biorefinery projects due to the scarcity of the tools and methodologies for assessment and evaluation of social aspects at present and the historically long detachment of the social sciences from the natural and engineering sciences (Lehmann et al. 2011). The indicators of social sustainability can be categorised as stand-alone social (sub-categorised into energy and food security; latter one being assessed by food price increase and sustainability factor), socio-economic, i.e. employment (including both generation of employment and requirement of labour), and socio-environmental (which include health as assessed by human exposure risk) (Parada et al. 2017). Food security is an important issue for the bioeconomy and in turn dependents upon land (Souza et al. 2015).

6 Guidelines for Sustainable Biorefinery

Guidelines for sustainable biorefineries can be proposed by adopting the similar strategies from other disciplines (such as chemical and engineering sciences) as detailed in Table 3. Such principles in chemical sciences can be stated as 'the design of chemicals and processes that reduce or eliminate the use or generation of hazardous substances' and green engineering concept as 'the design, commercialization, and use of processes and products in a way that minimises pollution, promotes sustainability, and protects human health without sacrificing economic viability and efficiency' (Gallego et al. 2011). These guidelines emphasise the sustainability element in biorefinery operations and may include several considerations for the development of a sustainable biorefinery as listed in Table 6.

7 Current Challenges and Future Prospects

Currently, biorefineries are being developed worldwide for sustainable synthesis of products and materials for various industrial sectors such as energy, transport, food, chemical, health, pharmaceutical, etc. One of the major future challenges for biorefinery is maintenance of socio-econo-environmental sustainability. For maximum utilisation of the complexity of lignocellulose, not a single technology or single biomass or production of a single product will be sufficient and obviously integration of different unit operations will be of utmost importance. The concept of biorefinery is already gaining popularity and few bio-based industries are currently operational or under demonstration stage. Recently, India has also mandated to establish various biorefineries in different parts of the country depending upon various technologies

Table 6 Some important considerations for the development of sustainable lignocellulosic biorefinery

Criteria	Important considerations[a]
Selection of biorefinery products	• High theoretical yield ($Y_{P/S}$) and large quantity of products • Ease of synthesis • One or two main products • One or two intermediate chemicals for revenue • Steam and electricity generated from process wastes
Feedstock availability, location and logistics	• Low-value and abundant agricultural waste non-competitiveness with feed/food and land use • Near to feedstock production zone and water supply • Provision of on-site short-term storage for product and feed chemicals • Material need to be rotated continuously • Possibility of nearby market
Biorefinery process	• Mild pretreatment with less degradation of carbohydrates • Specific pretreatment with respect to the desired range of products • Efficient fractionation of lignocellulosic components • Efficient microorganisms for better yields and productivities • Exploitation of both hexoses and pentoses • Ease of product recovery and purification • Recovery, reuse of water and reagents, if possible • Integration with power generation
Sustainability analysis	• Market analysis of products and by-products • System analysis based on social, environmental and economic aspects • Estimation of NPV and IRR • Operational and manufacturing capacity, manufacturing prices • Application of suitable software package • Economic feasibility and environmental performance should be considered • Eco-efficiency of biorefinery should be at par with the gasoline refinery

Table 7 Some operational lignocellulosic biorefinery industries

Industry and its location	Scale of operation (gallons/year)	Feedstock used	Conversion technology	Major product
Renewable Energy Institute, Ohio	PP (625,000)	Rice hulls and forest residues	Thermochemical gasification	Renewable diesel
INEOS New Planet Bioenergy LLC, Florida	Demo (8,000,000)	Municipal solid waste	Hybrid	Ethanol
RSA, Maine	Demo (1,500,000)	Forest resources	Biochemical	Biobutanol
Abengoa Kansas, United States	25,000,000+ 18 MW power	Wheat straw and agro-waste	Biochemical	Cellulosic ethanol and power
Beta Renewables, Crescentino, Italy	12,000,000	Agro-residues and energy crops	Biochemical	Cellulosic ethanol

and feedstocks. A brief overview of few lignocellulosic biorefineries currently operational with their scale of operation is provided in Table 7.

Currently, biorefinery industries face several challenges which can be broadly categorised into the biomass related and the process related challenges; whereas some are miscellaneous as they are common to both. Major challenges in each category are shown in Fig. 6. An overall challenging task is the commercial viability or economic sustainability of biorefineries. Cutting-edge research and state-of-the-art technologies need to be developed and implemented at all levels, i.e. lab-, pilot- and industrial production levels, which may require significant investments from government, academia and private industries. Consistent and dedicated research and development efforts are needed, especially for evaluation and validation of technologies being developed. There is a clear need for proper modelling, assessing and evaluating sustainability impacts on a life cycle-based analysis (https://www1.eere.energy.gov/bioenergy/pdfs/ibr_portfolio_overview.pdf; Agler et al. 2011).

8 Conclusion

In order to keep the bio-based economy sustainable, it is important to shift focus from the concept of single product from lignocellulosic feedstock to development of multitude of products. More technologies need to be developed, a range of products and co-products need to be enhanced and multiple feedstocks need to be utilised for better realisation of lignocellulosic biorefineries. Integration of various processes for conversion of various fractions of lignocellulosic biomass to different products is needed. In conclusion, there is an urgent need for proper and more systematic

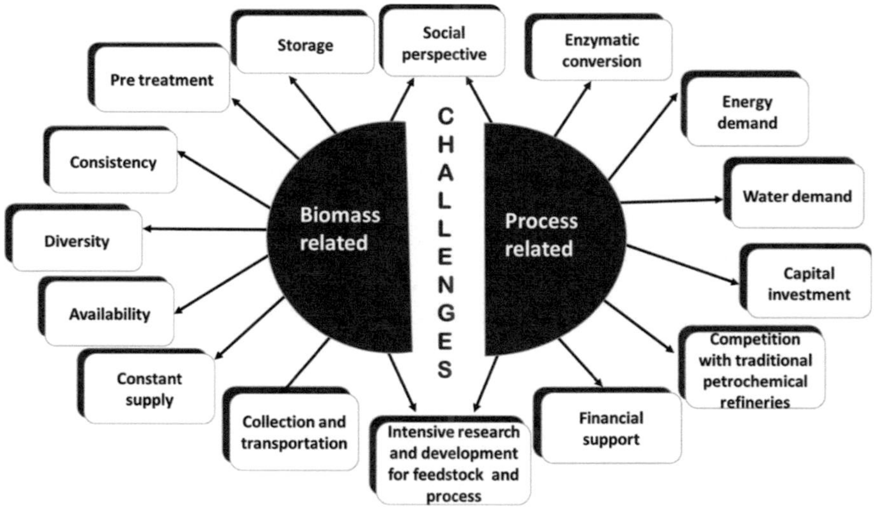

Fig. 6 Major current and emerging challenges for lignocellulosic biorefineries

improvement of feedstock, processes and the microbial and/or enzymatic performances for integration of biorefinery operations in a sustainable manner.

Acknowledgements The financial and infrastructural support from Central University of Haryana, Mahendergarh, Haryana, India is highly acknowledged. The authors also thank Science & Engineering research Board (SERB), Department of Science and Technology, Government of India for providing the financial support (SERB/LS/2016/00929).

References

Agarwal L, Dutt K, Meghwanshi GK, Saxena RK (2008) Anaerobic fermentative production of lactic acid using cheese whey and corn steep liquor. Biotech Lett 30(4):631–635

Agler MT, Wrenn BA, Zinder SH, Angenent LT (2011) Waste to bioproduct conversion with undefined mixed cultures: the carboxylate platform. Trends Biotechnol 29(2):70–78

Aithani D, Mohanty AK (2006) Value-added new materials from byproduct of corn based ethanol industries: blends of plasticized corn gluten meal and poly (ε-caprolactone). Ind Eng Chem Res 45(18):6147–6152

Al-Zuhair S (2007) Production of biodiesel: possibilities and challenges. Biofuels, Bioprod Biorefin 1(1):57–66

Amartey S, Jeffries TW (1994) Comparison of corn steep liquor with other nutrients in the fermentation of D-xylose by *Pichia stipitis* CBS 6054. Biotech Lett 16(2):211–214

Amidon TE, Liu S (2009) Water-based woody biorefinery. Biotechnol Adv 27(5):542–550

Anand P, Saxena RK (2012) A comparative study of solvent-assisted pretreatment of biodiesel derived crude glycerol on growth and 1, 3-propanediol production from *Citrobacter freundii*. New Biotechnol 29(2):199–205

Bakshi B (2002) A thermodynamic framework for ecologically conscious process systems engineering

Balat M (2008) Possible methods for hydrogen production. Energy Sources Part Recovery Util Environ Eff 31:39–50. https://doi.org/10.1080/15567030701468068

Barbirato F, Chedaille D, Bories A (1997) Propionic acid fermentation from glycerol: comparison with conventional substrates. Appl Microbiol Biotechnol 47(4):441–446

Behzadi S, Farid MM (2007) Examining the use of different feedstock for the production of biodiesel. Asia-Pac J Chem Eng 2(5):480–486

Berk Z (1992) Technology of production of edible flours and protein products from soybeans. FAO Agric Serv Bull 97:1–10

Bormann EJ, Roth M (1999) The production of polyhydroxybutyrate by *Methylobacterium rhodesianum* and *Ralstonia eutropha* in media containing glycerol and casein hydrolysates. Biotech Lett 21(12):1059–1063

Bos HL, Sanders JP (2013) Raw material demand and sourcing options for the development of a bio-based chemical industry in Europe. Biofuels, Bioprod Biorefin 7(3):246–259

BREW (2006) Medium and long-term opportunities and risks of the biotechnological production of bulk chemicals from renewable resources—the potential of white biotechnology. The BREW Project, Final Report, Utrecht, September 2006, 9–103

Brumbley SM, Purnell MP, Petrasovits LA, Nielsen LK, Twine PH (2007) Developing the sugarcane biofactory for high-value biomaterials. Int Sugar J 1297:5

Chandra R, Takeuchi H, Hasegawa T, Kumar R (2012) Improving biodegradability and biogas production of wheat straw substrates using sodium hydroxide and hydrothermal pretreatments. Energy 43:273–282. https://doi.org/10.1016/j.energy.2012.04.029

Cheng CL, Lo YC, Lee KS, Lee DJ, Lin CY, Chang JS (2011) Biohydrogen production from lignocellulosic feedstock. Biores Technol 102(18):8514–8523

Cherubini F, Ulgiati S (2010) Crop residues as raw materials for biorefinery systems—a LCA case study. Appl Energy 87(1):47–57

De Azeredo LAI, De Lima MB, Coelho RRR, Freire DMG (2006) A low-cost fermentation medium for thermophilic protease production by *Streptomyces* sp. 594 using feather meal and corn steep liquor. Curr Microbiol 53(4):335

De Bhowmick DG, Sarmah AK, Sen R (2018) Lignocellulosic biorefinery as a model for sustainable development of biofuels and value added products. Biores Technol 247:1144–1154

de Jong E, Higson A, Walsh P, Wellisch M (2012) Product developments in the bio-based chemicals arena. Biofuels Bioprod Biorefining 6:606–624. https://doi.org/10.1002/bbb.1360

Demirbas A (2005) Biodiesel production from vegetable oils via catalytic and non-catalytic supercritical methanol transesterification methods. Prog Energy Combust Sci 31(5):466–487

Doleyres Y, Beck P, Vollenweider S, Lacroix C (2005) Production of 3-hydroxypropionaldehyde using a two-step process with *Lactobacillus reuteri*. Appl Microbiol Biotechnol 68(4):467–474

Eggink G, Eenink AH, Huizing HJ (1994) PHB-producing microorganism and process for removing glycerol from a culture medium. Patent publication number WO9409146

Fahlén E, Ahlgren EO (2009) Assessment of integration of different biomass gasification alternatives in a district-heating system. Energy 34:2184–2195

Fan J, Shonnard DR, Kalnes TN, Johnsen PB, Rao S (2013) A life cycle assessment of pennycress (*Thlaspi arvense* L.)—derived jet fuel and diesel. Biomass Bioenergy 55:87–100

Gallego A, Hospido A, Moreira MT, Feijoo G (2011) Environmental assessment of dehydrated alfalfa production in Spain. Resour Conserv Recycl 55(11):1005–1012

Ganjyal GM, Reddy N, Yang YQ, Hanna MA (2004) Biodegradable packaging foams of starch acetate blended with corn stalk fibers. J Appl Polym Sci 93(6):2627–2633

Gennadios A (ed) (2002) Protein-based films and coatings. CRC Press

Gunasekara MY, Edwards DW (2003) Estimating the environmental impact of catastrophic chemical releases to the atmosphere: an index method for ranking alternative chemical process routes. Process Saf Environ Prot 81:463–474. https://doi.org/10.1205/095758203770866638

Hamelinck CN, van Hooijdonk G, Faaij AP (2005) Ethanol from lignocellulosic biomass: techno-economic performance in short-, middle- and long-term. Biomass Bioenergy 28:384–410. https://doi.org/10.1016/j.biombioe.2004.09.002

Hofer R, Bigorra J (2008) Biomass-based green chemistry: sustainable solutions for modern economies. Green Chem Lett Rev 1:79–97. https://doi.org/10.1080/17518250802342519
Hossain GS, Liu L, Du GC (2016) Industrial bioprocesses and the biorefinery concept. In: Larroche C Sanromán MA, Du G, Pandey A (eds) Current developments in biotechnology and bioengineering: bioprocesses, bioreactors and controls. Elsevier, 2017, pp 3–27. https://doi.org/https://doi.org/10.1016/b978-0-444-63663-8.00001-x
http://capitalenergy.biz/?p=6744 change to http://capitalenergy.biz/?p=6744. Accessed on 20 Dec 2017
https://www.aber.ac.uk/en/media/departmental/ibers/pdf/innovations/07/07ch8.pdf. Accessed on 20 Dec 2017
https://www1.eere.energy.gov/bioenergy/pdfs/ibr_portfolio_overview.pdf. Last accessed on 25 Dec 2017
Institute for Environment and Sustainability (2010) Analysis of existing environmental impact assessment methodologies for use in life cycle assessment. In: ILCD handbook, 1st ed. European Commission Joint Research Center
ISO (2006) ISO 14040:2006—environmental management—life cycle assessment—principles and framework. https://www.iso.org/standard/37456.html. Accessed 7 Dec 2017
Jerez A, Partal P, Martinez I, Gallegos C, Guerrero A (2005) Rheology and processing of gluten based bioplastics. Biochem Eng J 26(2):131–138
Kajaste R (2014) Chemicals from biomass–managing greenhouse gas emissions in biorefinery production chains–a review. J Clean Prod 75:1–10
Kamm B, Gruber PR, Kamm M (2007) Biorefineries—industrial processes and products. In: Ullmann's encyclopedia of industrial chemistry. Wiley-VCH Verlag GmbH & Co. KGaA
Kamm B, Kamm M (1997) Tagungsband/Symposium Chemie nachwachsender Rohstoffe: Wien, 9/10. Sept 1997. Bundesministerium für Umwelt, Jugend und Familie, Wien
Kamm B, Kamm M (2004) Principles of biorefineries. Appl Microbiol Biotechnol 64(2):137–145
Kemp R, Martens P (2007) Sustainable development: how to manage something that is subjective and never can be achieved? Sustain Sci Pract Policy 3:5–14. https://doi.org/10.1080/15487733.2007.11907997
Kılıc Apar D, Ozbek B (2007) Hydrolysis and solubilization of corn gluten by Neutrase. J Chem Technol Biotechnol 82(12):1107–1114
Kim JM, Whang JH, Kim KM, Koh JH, Suh HJ (2004) Preparation of corn gluten hydrolysate with angiotensin I converting enzyme inhibitory activity and its solubility and moisture sorption. Process Biochem 39(8):989–994
Kim J, Yun S, Ounaies Z (2006) Discovery of cellulose as a smart material. Macromolecules 39(12):4202–4206
Kumar R, Singh S, Singh OV (2008) Bioconversion of lignocellulosic biomass: biochemical and molecular perspectives. J Ind Microbiol Biotechnol 35:377–391
Lakaniemi AM, Tuovinen OH, Puhakka JA (2013) Anaerobic conversion of microalgal biomass to sustainable energy carriers–a review. Biores Technol 135:222–231
Langeveld H, Sanders J, Meeusen-van Onna MJG (2012) The biobased economy: biofuels, materials, and chemicals in the post oil era
Leduc S, Schwab D, Dotzauer E et al (2008) Optimal location of wood gasification plants for methanol production with heat recovery. Int J Energy Res 32:1080–1091. https://doi.org/10.1002/er.1446
Lehmann A, Russi D, Bala A et al (2011) Integration of social aspects in decision support, based on life cycle thinking. Sustainability 3:562–577. https://doi.org/10.3390/su3040562
Liu S, Lu H, Hu R, Shupe A, Lin L, Liang B (2012) A sustainable woody biomass biorefinery. Biotechnol Adv 30(4):785–810
Luo L, van der Voet E, Huppes G (2010) Biorefining of lignocellulosic feedstock–technical, economic and environmental considerations. Biores Technol 101(13):5023–5032
Maity SK (2015) Opportunities, recent trends and challenges of integrated biorefinery: part I. Renew Sustain Energy Rev 43:1427–1445. https://doi.org/10.1016/j.rser.2014.11.092

McCormick K, Kautto N (2013) The bioeconomy in Europe: an overview. Sustainability 5:2589–2608. https://doi.org/10.3390/su5062589
Menon V, Rao M (2012) Trends in bioconversion of lignocellulose: biofuels, platform chemicals & biorefinery concept. Prog Energy Combust Sci 38(4):522–550
Moncada BJ, Aristizábal MV, Cardona ACA (2016) Design strategies for sustainable biorefineries. Biochem Eng J 116:122–134. https://doi.org/10.1016/j.bej.2016.06.009
Moncada J, Cardona CA, Rincón LE (2015) Design and analysis of a second and third generation biorefinery: the case of castorbean and microalgae. Biores Technol 198:836–843
Mosier N, Wyman C, Dale B et al (2005) Features of promising technologies for pretreatment of lignocellulosic biomass. Bioresour Technol 96:673–686. https://doi.org/10.1016/j.biortech.2004.06.025
Nakamura CE, Whited GM (2003) Metabolic engineering for the microbial production of 1, 3-propanediol. Curr Opin Biotechnol 14(5):454–459
Nass LL, Pereira PAA, Ellis D (2007) Biofuels in Brazil: an overview. Crop Sci 47(6):2228–2237
Nel S (2010) The potential of biotechnology in the sugarcane industry: are you ready for the next evolution? Int Sugar J 112(1333):11
NREL Integrated Biorefinery Research Facility. Bioenergy. NREL. https://www.nrel.gov/bioenergy/ibrf.html. Accessed 31 Dec 2017
Posada JA, Osseweijer P (2016) Socioeconomic and environmental considerations for sustainable supply and fractionation of lignocellulosic biomass in a biorefinery context. In: Mussatto SI (ed) Biomass fractionation technologies for a lignocellulosic feedstock based biorefinery. Elsevier, Amsterdam, pp 611–631. https://doi.org/10.1016/B978-0-12-802323-5.00026-8
Palmqvist E, Hahn-Hägerdal B (2000) Fermentation of lignocellulosic hydrolysates. II: Inhibitors and mechanisms of inhibition
Parada PM, Osseweijer P, Posada Duque JA (2017) Sustainable biorefineries, an analysis of practices for incorporating sustainability in biorefinery design. Ind Crops Prod 106:105–123. https://doi.org/10.1016/j.indcrop.2016.08.052
Quéméneur M, Hamelin J, Barakat A et al (2012) Inhibition of fermentative hydrogen production by lignocellulose-derived compounds in mixed cultures. Int J Hydrog Energy 37:3150–3159. https://doi.org/10.1016/j.ijhydene.2011.11.033
Samarasinghe S, Easteal AJ, Edmonds NR (2008) Biodegradable plastic composites from corn gluten meal. Polym Int 57(2):359–364
Santoyo-Castelazo E, Azapagic A (2014) Sustainability assessment of energy systems: integrating environmental, economic and social aspects. J Clean Prod 80:119–138
Sarma SJ, Ayadi M, Brar SK (2017) Biorefinery: general overview. In: Platform chemical biorefinery, pp 21–32
Seider W, Seader J, lewin D, Widadgo S (2010) Product and process design principles. Synthesis, analysis and evaluation, 3rd ed. Wiley
Shonnard DR, Hiew DS (2000) Comparative environmental assessments of VOC recovery and recycle design alternatives for a gaseous waste stream. Environ Sci Technol 34:5222–5228. https://doi.org/10.1021/es0010857
Shukla R, Cheryan M (2001) Zein: the industrial protein from corn. Ind Crops Prod 13(3):171–192
Souza GM, Victoria RL, Joly CA, Verdade LM (2015) Bioenergy & sustainability: bridging the gaps. SCOPE, Sao Paulo
Speight JG (2014) The chemistry and technology of petroleum. CRC press
Stefanis SK, Livingston AG, Pistikopoulos EN (1995) Minimizing the environmental impact of process plants: a process systems methodology. Comput Chem Eng 19:39–44
Szuhaj BF (ed) (1989) Lecithins: sources, manufacture & uses, vol 12. The American Oil Chemists Society
Talebian-Kiakalaieh A, Amin NAS, Hezaveh H (2014) Glycerol for renewable acrolein production by catalytic dehydration. Renew Sustain Energy Rev 40:28–59
Tan EC, Talmadge M, Dutta A, Hensley J, Snowden-Swan LJ, Humbird D, Schaidle J, Biddy M (2016) Conceptual process design and economics for the production of high-octane gasoline

blendstock via indirect liquefaction of biomass through methanol/dimethyl ether intermediates. Biofuels, Bioprod Biorefin 10(1):17–35

Vancauwenberge JE, Slininger PJ, Bothast RJ (1990) Bacterial conversion of glycerol to beta-hydroxypropionaldehyde. Appl Environ Microbiol 56(2):329–332

Wetterlund E, Soderstrom M (2010) Biomass gasification in district heating systems—the effect of economic energy policies. Appl Energy 87:2914–2922

World Commission on Environment and Development (1987) Report of the World Commission on environment and development: our common future. available at http://www.un-documents.net/our-common-future.pdf. Accessed on 20 Dec 2017

Yao K, Tang C (2013) Controlled polymerization of next-generation renewable monomers and beyond. Macromolecules 46(5):1689–1712

Young DM, Cabezas H (1999) Designing sustainable processes with simulation: the waste reduction (WAR) algorithm. Comput Chem Eng 23:1477–1491. https://doi.org/10.1016/S0098-1354(99)00306-3

Zheng Y, Pan Z, Zhang R (2009) Overview of biomass pretreatment for cellulosic ethanol production. Int J Agric Biol Eng 2:51–68

Chemicals and Fuels Production from Agro Residues: A Biorefinery Approach

Desikan Ramesh, Iniya Kumar Muniraj, Kiruthika Thangavelu and Subburamu Karthikeyan

Abstract Alternative fuel production technology in order to combat the present scenario of climate change issues has been in the transition stage of generating high-value low volume chemicals, fuels and low-value high volume bulk products (biofuels). This concept of biorefinery is the future of biomass processing technologies where complete utilization of biomass and zero release of waste can be achieved. The agro residue biorefinery aims in sustainable approach which provides a good solution for sustainable ways of utilizing agricultural residues. Agro residues are by-products of agricultural crop production and processing, which are abundantly available at lower price. Agro residues are one of the major resources of unexploited potential lignocellulosic feedstocks. It includes straws, leaves and plant materials left in the field after harvesting of the crop. Its characteristics would vary with crops, species and environmental conditions. Annual agro residue production potential in India is ca. 550 MT. Currently, most of the residues are underutilized or burnt in situ, creating serious environmental pollutions. In order to utilize and effective disposal of these wastes, several methods are tried for tapping the energy/bioproducts from various crop residues via biochemical or thermochemical conversion routes. Biorefinery technologies can offer a platform for production of high-value chemicals and fuels from these residues, which are value-added products as well as provide more income for agriculturists. This chapter aims in bringing out sources of agro residues, the current state of the art of biomass processing and conversion viable technologies, and recent developments in the biorefinery of agro residues, and finally sheds light on commercialization of agro residue biorefinery.

D. Ramesh (✉)
Horticultural College & Research Institute for Women, Tamil Nadu Agricultural University, Tiruchirappalli, India
e-mail: rameshd@tnau.ac.in

I. K. Muniraj
Kumaraguru Institute of Agriculture, Sakthinagar, Erode, India

K. Thangavelu · S. Karthikeyan
Agricultural Engineering College and Research Institute, Tamil Nadu Agricultural University, Coimbatore, India

N. Srivastava et al. (eds.), *Sustainable Approaches for Biofuels Production Technologies*, Biofuel and Biorefinery Technologies 7,
https://doi.org/10.1007/978-3-319-94797-6_3

Keywords Biorefinery · Biochemicals · Thermochemical conversion technologies · Paddy straw

1 Introduction

Petroleum products play a major role in our modern life, and its continued usage has created negative impacts on the environment. Fossil fuels-based transport sector is the best example of negative impacts on the environment such as air pollution, climate change and global warming, and also affects nation's energy security and economy. Apart from this, emissions from the in situ burning of biomass stubble are the other causes to increase the CO_2 level and also air/land pollution. According to the World Energy Outlook 2002 (IEA 2002), the per capita emissions of OECD and transition economies are projected to reach 13 tonnes and 11 tonnes, respectively, in 2030. Renewable energy sources are mainly focused on alternatives to fossil fuels and their associated products due to its renewability, plenty available and also low cost. Population growth and unequal social development have exacerbated the vulnerability of our societies to the fragility of the world's climate system and the impacts of natural events. At this juncture, attention is naturally focused on biomass as energy resource looking for alternate energy sources, which can make a significant contribution to satisfy the energy needs of society with environmental friendly (Tao et al. 2013). Among the different biomass resources, lignocellulosic biomass feedstocks are potential candidates for promoting the transition from the petroleum-based economy to bioeconomy of the country for a sustainable development, and it reduces the depends on foreign oil imports and money (Raman et al. 2015; Xu et al. 2016). Agro residues are the main resource that falls under lignocellulosic biomass feedstocks category. It was obtained as by-products from agricultural crop production and processing operations. It includes straws, leaves and plant materials left in the field after harvesting of the crop. Its characteristics would vary with crops, species and environmental conditions (De Bhowmick et al. 2018). This residues are one of the unexploited potential resources, which offer a new business platform for production of variety of biofuels, fine chemicals and value-added products to displacing petroleum-based products. An estimation shows that theoretical energy potential of global agricultural resides produced in a year is ranged from 14.6 to 123 EJ (WBA Global Bioenergy Statistics 2017). Effective utilization of these residues is currently a big challenge for industries involved in production of biofuels and biochemical/bioproducts. Bioethanol production from this feedstock consumes more energy and processing cost than that of first-generation biofuel crops. Production of multiple products from agro residues is a viable solution to minimize the biofuel's price to compete with conventional fuels and also minimize the waste generation in each bioprocessing of biofuel production. Similar to a petroleum refinery, the multiple products from these wastes can be achieved via selected processes of different biomass conversion technologies.

Biorefinery mainly focused on bioenergy as primary product and chemicals as secondary products from biomass resources and vice versa. The biorefinery is a good option for effective utilization of renewable feedstocks for multiple products generation (bioenergy, biofuel and fine biochemicals) and minimizes the huge quantity of waste (Ahring and Westermann 2007; Cabeza et al. 2016; Christopher et al. 2017; Dominguez et al. 2014). This will help in safe handling and disposal of bulk quantity of wastes and lower the pollution in the environment. In other words, the by-products of the first process would be used as substrate/raw materials for next product production. For example, the biorefinery approach is applied to sugarcane crop to obtain bioethanol and fine chemicals. So far, four generations are developed for biorefineries, viz. first-generation (food crops or animal fat), second-generation (nonfood crops), third-generation (mix of different biomass feedstocks) and fourth-generation (vegetable oil). Among them, first-generation biorefineries are commercialized in different countries. Biorefinery offers energy and value-added products as well as provide more income to agriculturists (Zhang et al. 2016).

In this chapter, the different types of agro residues and their sources, viable conversion technologies, biorefinery approaches, constraints in commercialization of biorefineries, and paddy straw as a potential source for biorefinery are briefly discussed.

2 Agro Residues and Resources

Agro residues are obtained after harvesting or processing or both operations of agricultural crops. It may be in the form of solid or semi-solid, and it depends upon the agro products obtained. Generally, the agro residues can be classified into primary and secondary residues (Fig. 1). In case of primary residues, the residues are collected in the field itself after the harvesting of main agro product from crop, e.g. plant materials, stalks, leaves, etc. The secondary residues are generated from during the processing of main agricultural produce, e.g. rice husk, maize cob, etc. Residue-to-product ratio (RPR) is an important terminology related to residue calculation for a particular crop, and the value of RPR varies from crop to crop. RPR is defined as the ratio between the weight of the crop residue to the total weight of agro product yield for the selected crop. The range values of RPR for primary and secondary agro residues of different agricultural crops are 0.05–4.00 and 0.15–2.00, respectively (Table 1). RPR value is more useful to predict the quantity of agro residue generated for the selected crop. Multiplication of crop yield with RPR value would give waste generation for the calculated crop and RPR value also used in biomass assessment studies conducted for a location/state/country. Theoretical estimation of agro residues production per year for India and world is 500–550 MT and 3.6–17.2 billion tonnes (IARI 2012; WBA Global Bioenergy Statistics 2017). A huge amount of these wastes should be used effectively for multiple products productions via suitable biorefinery approach.

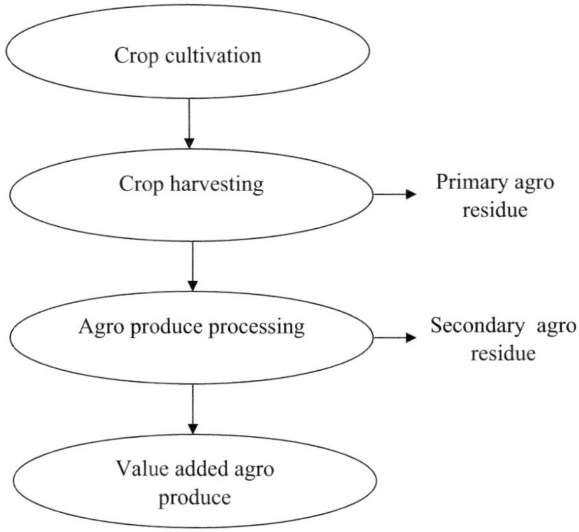

Fig. 1 Illustration of primary and secondary agro residue generated in harvesting and processing of agricultural crops

2.1 Availability of Agro Residues

During the processing of agricultural crops such as paddy, wheat, sugarcane and maize, a copious amount of residues are generated, for instance, 1 tonne of sugarcane generates 300 kg of bagasse, 1 tonne of paddy generates 0.75 tonnes of straw and the equal amount of stover is generated while processing corn. All these residues are rich in sugars and can be utilized for the biorefinery. A significant quantity of the residues will be used for fodder, manure and local use for trashing house. Globally, various countries generate agricultural residues in enormous amounts, and their availability is depicted in Fig. 2. India achieved self-sustainability in food grain production, and agriculture is main resource income for the farming community in most of the rural villages. Eight crops, viz. rice, wheat, bajra, jowar, sugarcane, cotton, groundnut and oilseeds, are majorly cultivated in 11 different states of India. The details of annual crop residues generated, surplus availability and their power generation potential are presented in Table 2. An estimate shows that annual power potential from agro residues alone for India is calculated as 18729.9 MWe from 511 MT of wastes.

Uttar Pradesh and Punjab are leading states in major agro residues production from both wheat and paddy crops. Annual agro residue production, major crops and types of residues in India are presented in Table 3. Among the crops, sugarcane, rice and wheat crops generate major share for agro residues generation in India. Sugarcane crop production and processing can contribute an amount of 2,76,250 metric tonnes of primary and secondary residues and paddy stands as a second largest contributor in our country.

Table 1 The RPR ratio for different agricultural crops

Crop group	Crop	Residue		RPR	References
Cereals	Rice	Primary	Straw	1.50	Hiloidhari and Baruah (2011a, b)
		Secondary	Husk	0.20	Singh et al. (2008a, b, c)
	Wheat	Primary	Stalk	1.50	
		Secondary	Pod	0.30	
	Maize	Primary	Stalk	2.00	
		Secondary	Cob	0.30	
	Bajra	Primary	Stalk	2.00	Friedl et al. (2005)
		Secondary	Cob	0.33	Hiloidhari et al. (2014)
			Husk	0.30	Raveendran et al. (1995)
	Barley	Primary	Straw	1.30	Friedl et al. (2005)
	Jowar	Primary	Stalk	1.70	
		Secondary	Cob	0.50	Hiloidhari et al. (2014)
			Husk	0.20	Raveendran et al. (1995)
Millets	Small millet	Primary	Straw	1.20	Friedl et al. (2005)
	Ragi		Straw	1.30	
	Kodo millet		Stalk	1.16	Biomass Knowledge Portal
Oilseeds	Mustard and rapeseed	Primary	Stalk	1.80	Singh et al. (2008a, b, c)
	Sesame		Stalk	1.20	Zabaniotou et al. (2008)
	Linseed		Stalk	1.47	Hiloidhari et al. (2014)
	Niger		Stalk	1.00	
	Safflower		Stalk	3.00	
	Soybean		Stalk	1.70	Kis et al. (2009)
	Groundnut		Stalk	2.00	Jekayinfa and Scholz (2009)
		Secondary	Shell	0.30	
	Sunflower	Primary	Stalk	3.00	Zabaniotou et al. (2008)
Pulses and legumes	Tur(arhar)	Primary	Stalk	2.50	Singh et al. (2008a, b, c)
		Secondary	Husk	0.30	Biomass Knowledge Portal
	Avare	Primary	Stalk	1.10	
	Lentil		Stalk	1.80	Hiloidhari et al. (2014)
	Guar		Stalk	2.00	Singh et al. (2008a, b, c)
	Green gram		Stalk	1.10	

(continued)

Table 1 (continued)

Crop group	Crop	Residue		RPR	References
		Secondary	Husk	0.15	Biomass Knowledge Portal
	Horse gram	Primary	Stalk	1.30	
	Red gram		Stalk	1.10	
	Moth bean		Stalk	1.80	
	Peas and beans		Stalk	0.50	
Sugar crop	Sugarcane	Primary	Top and leaves	0.05	Singh et al. (2008a, b, c)
		Secondary	Bagasse	0.33	
Horticulture	Banana	Secondary	Peel	3.00	Wilaipon (2009)
	Coconut	Primary	Frond	4.00	Rahman (2006)
		Secondary	Husk and pith	0.53	Minowa et al. (1998)
			Shell	0.22	Biomass Knowledge Portal
	Areca nut	Primary	Frond	3.00	Hiloidhari et al. (2014)
		Secondary	Husk	0.80	Pilon (2007)
Fibres	Cotton	Primary	Stalk	3.80	Jekayinfa and Scholz (2009)
		Secondary	Husk	1.10	Hiloidhari et al. (2014)
			Boll shell	1.10	Caglar and Demirbas (2001)
	Jute	Primary	Stalk	2.00	Asadullah et al. (2008)
Spices and condi-ments	Cardamom	Primary	Stalk	0.64	Biomass Knowledge Portal
	Coriander		Stalk	1.15	
	Cumin seed		Stalk	1.55	
	Dry chilly		Stalk	1.50	
	Turmeric		Stalk	0.30	
Plantains	Coffee	Primary	Pruning and Wastes	4.00	
		Secondary	Husk	0.50	
	Tea	Primary	Sticks	1.00	
	Rubber	Primary	Wood	3.00	
		Secondary	Wood	2.00	
	Tobacco	Primary	Stalk	1.00	
Vegetables	Onion	Primary	Stalk	0.05	

(continued)

Table 1 (continued)

Crop group	Crop	Residue		RPR	References
	Dry ginger		Stalk	0.05	
	Garlic		Sheath	0.25	
			Stalk	0.05	
Tubers	Potato	Primary	Leaves	0.76	
			Stalk	0.05	
	Sweet potato		Stalk	0.10	
	Tapioca		Stalk	0.75	
Medicinal	Isabgol	Primary	Stalk	1.10	

Source production data were taken from http://faostat3.fao.org/browse/Q/QC/E and residues values were obtained manual conversion ratio from each crop yield

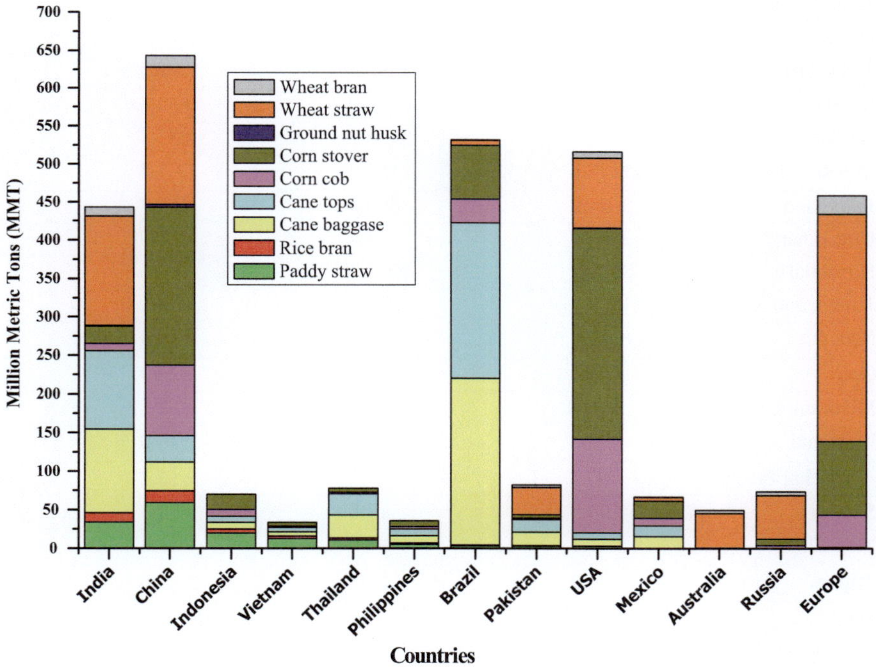

Fig. 2 Total agricultural residues generated by various countries in 2012

Table 2 Estimated annual agro residues production, surplus availability and their power production potential in India

State	Agro residue production (kT/yr)	Surplus (kT/yr)	Power potential (MWe)
Uttar Pradesh	60322.2	13753.7	1748.3
Punjab	50847.6	24843.0	3172.1
Maharashtra	47624.8	14789.9	1983.7
West Bengal	35989.9	4301.5	529.2
Karnataka	34167.3	9027.3	1195.9
Madhya Pradesh	33344.8	10329.2	1373.3
Rajasthan	29851.3	8645.6	1126.7
Haryana	29034.7	11343.0	1456.9
Gujarat	29001.0	9058.3	1224.8
Bihar	25756.9	5147.2	640.9
Andhra Pradesh	24871.7	4259.4	520.8
Tamil Nadu	22507.6	8899.9	1159.8
Odisha	20069.5	3676.7	429.1
Telangana	19021.5	2697.2	342.5
Kerala	11644.3	6351.9	864.4
Assam	11443.6	2436.7	283.7
Chhattisgarh	11272.8	2127.9	248.3
Jharkhand	3644.9	890.0	106.7
Uttarakhand	2903.2	638.4	81.0
Himachal Pradesh	2896.9	1034.7	132.6
Jammu and Kashmir	1591.3	279.5	37.1
Manipur	909.4	114.4	14.3
Goa	668.5	161.4	20.9
Mizoram	511.1	8.5	1.1
Nagaland	492.2	85.2	10.0
Arunachal Pradesh	400.4	74.5	9.2
Sikkim	149.5	17.8	2.3
Meghalaya	61.1	91.6	11.3
Tripura	40.9	21.3	3.0
Total	511040.9	145105.7	18729.9

Source http://biomasspower.gov.in/ (as per May 20, 2016)

Table 3 Details of annual residues production from main crops in India (Thomas et al. 2017)

Crop	Annual production, metric tonnes	Types of agro residues generated
Sugarcane	2,76,250	Bagasse, top and leaves
Rice	1,45,050	Stalks, straw
Wheat	78,000	Pods, stalks
Banana	80,000	Residue, cobs
Maize	18,000	Stalks, fronds
Coconut	13,125	Husk and pith, shell
Millets	12,410	Stalks, cobs
Bajra	7690	Stalks, husks
Cassava	6060	Solid waste, starch from roots
Arhar	1950	Husks, stalks

3 Biorefinery

Biorefinery involves several sequences of operations to disintegrate/convert the biomass into different bioproducts (biochemicals/biofuels/biomaterials) and bioenergy, in other words, conversion of biomass into useful bioproducts via efficient biomass conversion technologies with minimal waste generation. The biomass conversion technologies may be based on thermochemical, chemical or biochemical conversion routes and/or their combinations. The biorefinery is a facility that integrates biomass conversion process and equipment to produce fuels, power and chemicals from biomass (NREL 2009). In the biorefinery, transformation of recovered sugars from agro residues into fuels and chemicals are achieved by combinaion of new fermentation and thermochemical processes. Biorefinery is a clear example of industrial symbiosis, as it involves careful management and utilization of materials, products and wastes in a desirable way. There are three phases of biorefinery known, viz., phases I, II and III (Octave and Thomas 2009).

3.1 Phase I Biorefinery

The phase I biorefinery utilizes grain as feedstocks such as corn and wheat. The main difference in phase I and phase II biorefinery is that it has fixed processing capabilities and produces a fixed amount of ethanol and other feed products.

3.2 Phase II Biorefinery

It has more flexibility than phase I, wherein it can produce more end products and far more flexibilities. Examples of corn dry milling and wet milling are phase I and phase II biorefinery, respectively. Under phase II biorefinery if corn is used as feedstock, it can produce multiple products such as gluten feed, high fructose and corn syrup besides starch, glucose and dextrose, ethanol, gluten meal and corn oil.

3.3 Phase III Biorefinery

Recently, phase III biorefinery gains attention; it combines a mix of biomass feedstocks and yields an array of products by employing the combination of different technologies. Although it is in the developing stage, phase III system offers more advantages than other phases. It can simultaneously operate wet and dry biomasses both treated and untreated, recovered sugars such as cellulose, hemicellulose would be combined or processed in batch wise and high-value chemicals would be generated. Lignin would be used in direct combustion to generate steam and electricity.

3.4 Types of Biorefinery

The general scheme on biorefinery initially starts with separation of plant components by grinding followed by a fractionation by biological and physicochemical technologies. This enables the role of biomass extracts to be used as functional compounds. The next step is synthesis of agro-industrial products and development of a large number of bio-based products. Classification of biorefineries based on different types of feedstock used is shown in Fig. 3.

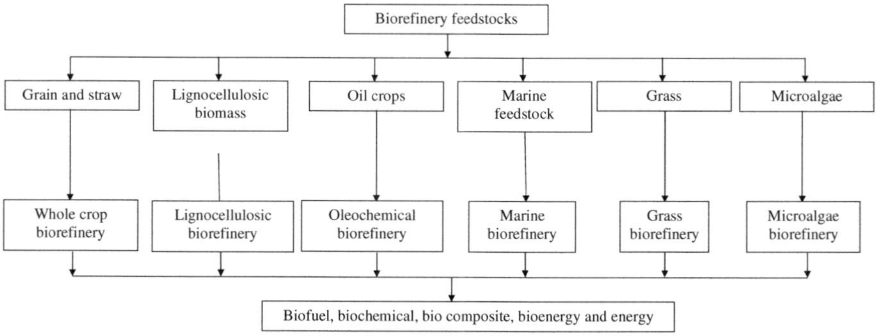

Fig. 3 Biorefinery technologies used for different feedstocks

4 Important Energy and Chemicals Recovery Routes Under Biorefinery Approach for Agro Residues

Biomass energy conversion technologies are mainly divided into three categories, viz. thermochemical, biochemical and chemical conversion technologies to obtain different types of biofuels, bioenergy and fine chemicals from biomass feedstock. Among the four different biofuel generation technologies, first-generation biofuel is successfully commercialized in different countries. Other biofuel generation technologies are still in infancy stage to address the challenges involved in optimizing process conditions and also reduce the processing cost. The additional amount spent on 2, 3 and 4G technologies for biofuel production should have an impact on fuel price and difficult to compete with conventional fossil fuels.

4.1 Anaerobic Digestion

A microbial consortium is involved in biodegradation of organic matter in four different consecutive steps under anaerobic conditions to yield methane-rich gas. Anaerobic digestion is also called as biomethanation process. Two end products obtained from anaerobic digestion process are gaseous biofuel and biodigested slurry. The gaseous fuel is referred as biogas, and it can be used for thermal applications, lighting and engine running. Most of the agro residues are suitable to produce biogas via anaerobic digestion process (Adney et al. 1991).

4.2 Gasification

It is one of the thermochemical conversion technologies used to convert the biomass feedstock into gaseous biofuel. Gasification involves thermal cracking and incomplete combustion of biomass under a limited amount of air/O_2 supply. Gaseous fuel generated at the end of the gasification process. Depending on the oxidizing agents used in gasification, gaseous fuel is referred as producer gas for air supply and syngas for O_2 supplied. The gas compositions of the producer gas are % carbon monoxide, % hydrogen and traces of methane. The biomass gasifier is a device used for this purpose. Based on the flow of oxidizing agent and feedstock materials, the biomass gasifier is mainly classified into three categories, viz. downdraft, updraft and cross draft gasifier. For gasifying of biomass with higher moisture content, supercritical water or plasma gasification can be used. Deterrents of wet biomass gasification are the higher investment, more energy and inputs required and the skilled technical person required to operate this equipment (Chen et al. 2015).

4.3 Pyrolysis

Thermal degradation of biomass feedstocks occurred in between 350 and 500 °C under the absence of air or O_2. The end products of this process may be in the form of gaseous fuel or solid fuel or liquid biofuel. The choice of end product mainly depended on the reaction temperature, heating rate, particle size and reaction time (Chen et al. 2017).

4.4 Combustion

Combustion is the process of burning biomass feedstocks under excessive aerated environment to generate the heat energy. Several factors that have an influence on heat release during combustion process are biomass composition, biomass types and plant age. Generally, combustion is used in the last stage of biorefinery approaches for converting entire biomass into heat energy for electricity production and ashes. In situ burning of agricultural residues is not a new technique, and farmers adopting this technique for quick and easy disposal of agro residues for land preparation for next crop cultivation in the consecutive season (Byun and Han 2016; Eynde et al. 2016; Hellier et al. 2015).

4.5 Biochemicals

Effective utilization of agro residues would lead to additional income, low waste generation and reduced dependence on other conventional resources (Beller et al. 2015). The biochemicals can be derived from different components of residues such as cellulose, hemicellulose and lignin. Sugars in the form of polysaccharide both structural and storage are the important part of the plant. However, in general, the sugars are always in close association with lignin. The biorefinery would first separate the sugar component of the plant without much damage and further remove the other components. Among the sugars, simple sugars like glucose and hexoses are the predominant compounds for many applications like ethanol for biofuels. The derived product from glucose is a lactic acid, which serves as a basic molecule in chemistry (Bouaid et al. 2010). For example, lactate esters are used as green solvents in industries. Further transformation of lactic acid into high-value chemicals such as acrylic acid and 1,2 propane diol would also be possible. Succinic acid is one of the important derivates of glucose that can also be produced by chemical inducers.

5 Selection of Biorefinery Technologies

Biochemical composition of agro residues would vary from crop to crop, genotypes, season to season, soil and environmental conditions. Several types of biomass conversion technologies are employed in the biorefinery approaches. The selection of appropriate technologies is strongly based on feedstock composition and the focused end products. The biochemical composition of different agro residues is presented in Table 4. Three major components of this kind of biomass are varied for different resources, and values of cellulose, hemicellulose and lignin are in the range of 10–53%, 0.15–65% and 5–45%, respectively. For example, an agro residue with higher carbohydrate content is suitable for biochemical conversion route to produce bioalcohols from fermentable sugars, whereas biomass composition is not a major issue for thermochemical conversion route (Henry 2018). And also, the biomass with higher moisture (more than 50%) would not suitable for most of the thermochemical conversion methods such as gasification and combustion process. Selection criteria are based on raw materials, end products, technologies used, processing cost and the market value of end products. If energy production is focussed, combustion is the best route among the thermochemical conversion technologies. Benefits of combustion technologies are less operational costs and higher energy output than other technologies. A rough estimate showed that production of biochemical, biofuels and energy generation from biomass may utilize 20, 40, and 40% of biomass used for the process (de Jong and Jungmeier 2015). The biorefinery technology is classified based on the number of feedstocks used, process involved and end products produced.

6 Constraints in Commercialization

The sustainability of biorefinery industries is depending on efficient conversion technologies, government policies, incentives and also techno-economical viable process availability. Even though each technology has its own merits and demerits, there are several biomass power plants in India shut down their operations. The reason is non-availability of feedstock, higher feedstocks price and fewer incentives. In order to avoid this kind of situation for agro residue-based biorefineries, the following points should be considered before the biorefinery plant installation:

(a) Survey on crop residue assessment,
(b) Feedstock collection via contract farming,
(c) Policy for promoting biorefinery and
(d) Good supply chain and logistics.

Table 4 Biochemical composition of different agro residues

Agro residues	Cellulose (%)	Hemicellulose (%)	Lignin (%)	References
Rice straw	28.1–43.77	20.47–31.42	4.84–23.3	Chen et al. (2011a, b), ECN—Phyllis2, Prassad et al. (2007a, b), Rai et al. (1989), Sarnklong et al. (2010)
Wheat straw	28.8–51.5	10.5–43	5.4–30	ECN—Phyllis2, Gao et al. (2016), Mani et al. (2006a, b), Esteghlalian et al. (1997), Motte et al. (2014)
Almond shell	29.0–31.1	28.0–38.0	27.7–35	Dhyani and Bhaskar (2017)
Barley straw	31–45	21.9–38	6.3–19	Saini et al. (2015), Mani et al. (2006a, b), Nigam et al. (2009), Cai et al. (2017)
Cashew nut shell	41.3	18.6	40.1	Dhyani and Bhaskar (2017)
Coir pith	36–43	0.15–0.25	41–45	Saini et al. (2015)
Corn straw	27.9–42.6	14.8–21.3	8.2–19	Diaz et al. (2015), Bilal et al. (2017a, b)
Eucalyptus	45–51	11–18	29	Bilal et al. (2017a, b)
Flax straw	36.7	34.4	28.9	Dhyani and Bhaskar (2017)
Groundnut shell	35.7	18.7	30.2	Dhyani and Bhaskar (2017)
Horticultural waste	34.5	28.6	36	Bilal et al. (2017a, b)
Jute fibres	45–53	18–21	21–26	Bilal et al. (2017a, b)
Millet husk	33.3	26.9	14	Dhyani and Bhaskar (2017)
Nut shells	25–30	25–30	30–40	Bilal et al. (2017a, b)
Oat straw	31–39.4	27–38	16–19	Nigam et al. (2009), Sanchez (2009)
Palm fibre	35.4	19.9	27.3	Laghari et al. (2016)
Rice husk	25–44.12	12.0–29.3	7.28–31	Nordin et al. (2007), Ludueña et al. (2011), Wang et al. (2012), Braga et al. (2013), Cai et al. (2017)
Rice straw	28–37.81	22.3–28	12–19	Prassad et al. (2007a, b), Chen et al. (2011a, b), Phan et al. (2014), Saini et al. (2015), Cai et al. (2017)
Rye straw	30.9–35	21.5–30	16–25.3	Sun and Cheng (2005), Garcia-Cubero et al. (2009), Sanchez (2009)

(continued)

Table 4 (continued)

Agro residues	Cellulose (%)	Hemicellulose (%)	Lignin (%)	References
Sorghum straw	2.0–35.0	24.0–27.0	15.0–21.0	Cai et al. (2017)
Sorted refuse	60	20	20	Bilal et al. (2017a, b)
Tamarind kernel	10–15	55–65		Bilal et al. (2017a, b)
Tobacco stalk	42.4	28.2	27	Dhyani and Bhaskar (2017)
Wheat bran	10.5–14.8	35.5–39.2	8.3–12.5	Bilal et al. (2017a, b)
Wheat shell	10–15	30	4–8	Bertero et al. (2012)
Wheat straw	29–49	22.3–50	5–21	Mani et al. (2006a, b), McKendry (2002), Ballesteros et al. (2006), Butler et al. (2013), Saini et al. (2015), Bharathiraja et al. (2017), Cai et al. (2017), Gaurava et al. (2017)

6.1 Survey on Crop Residue Assessment

A survey is compulsory to know about the availability, surplus, price, bulk straw handling machinery and current disposal methods for straw management. It also gave a clear-cut idea about access points for straw collection and work out estimate for the cost involved for feedstock supply and logistics. The details may be collected by interview methods and compiled to generate the biomass database for the survey region.

6.2 Feedstock Collection via Contract Farming

In order to obtain the agro residues in continuous manner and control the feedstock price, contract farming may be followed. Sugar industry is the best example for contract farming to get the assured feedstocks collected from the contracted farmers. This kind of arrangement would help to run the sugar industry in continuous manner and to sustain the market by supplying the sugar to their consumers.

6.3 Policy for Promoting Biorefinery

Straw burning in the field would lead to air/land pollution and create a negative impact on the environment. Generally, countries announced that straw burning is illegal and implemented strict laws to prohibit the straw burning. Recently, Indian government formulated a national policy for management of crop residues (NPMCR) and implemented to stop the burning residues in the field and to promote different uses of their residues to make bequests or pellets for industrial uses (State of Indian Agriculture 2015).

6.4 Good Supply Chain and Logistics

This stage is involved in harvest and transport the straw from the paddy field to biorefinery unit, since the straw belongs to low bulk density materials, which occupies more space for storage and also involves higher transport cost. To address these issues, the harvested straw may be baled and stocked in the field. The moisture content of the feedstock is not a problem when it is used for biomass pretreatment. In order to store it in the continued area, straw may be powdered and then briquetted. This will reduce storage space, transport cost and also ease in handling the materials. The argument for the higher cost involved for briquetting process may be overruled by the cost involved for raw material handling both transport and storage area.

7 Case Study: Paddy Straw a Potential Feedstock for Biorefinery

In paddy processing, straw left in the field after harvesting of paddy grain is referred as primary agro residue. Paddy husk is obtained in the further processing of paddy grains to rice, and this husk is referred as secondary agro residues (Fig. 4). This straw contains three major components, viz. cellulose, hemicellulose and lignin. It is the largest source of lignocellulosic feedstock generated from the paddy crop.

Global paddy straw production per year is about 731 million tonnes with a major share of 91.30% by Asia (Binod et al. 2010).

Theoretical estimation of paddy straw and rice husk produced in India is about 157 and 21 million tonnes, respectively. Presently, the paddy straw has wide applications such as fodder, manure, roofing materials, fiberboard, etc. (Fig. 5). Paddy husk also often used to generate the heat energy for thermal applications. Burning/combustion is the most popular method for quick disposal of the paddy straw. According to Singh et al. (2008a, b, c), paddy straw production is ca. 17 million tonnes in Punjab alone, and in situ burning is practiced for 90% of this waste. The two in situ burning methods of the straw at site practicing by farmers are partial burning and complete burning. Farmers preferred to adopt complete burning at the agricultural field for low-cost quick disposal method. Burning is a temporary solution for disposal of huge amount of paddy straw. It simultaneously creates health problems due to worst air quality and also destroying on microorganisms in the soil. The problems associated with the

Fig. 4 Primary (paddy straw) and secondary agro residues (paddy husks) produced in harvesting and processing of paddy crop

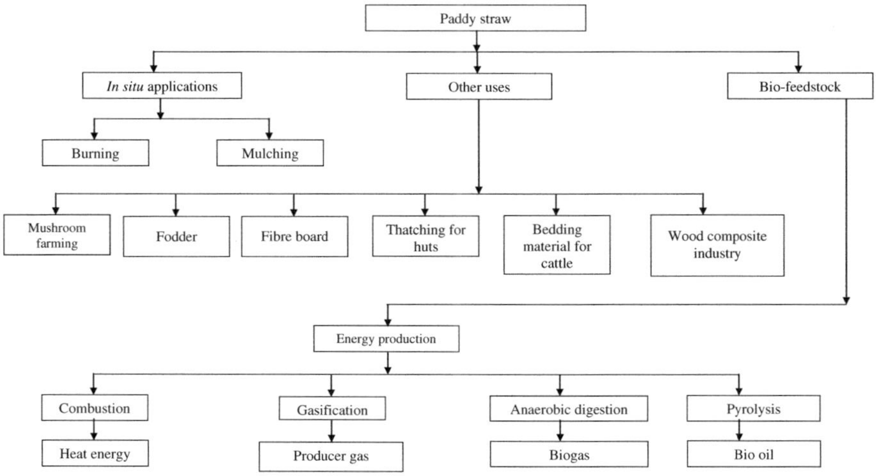

Fig. 5 Conventional disposal and energy recovery methods used for the paddy straw

handling of paddy straw are as follows: (i) transportation is a big issue, (ii) requires more space for transport and storage due to low bulk density (Fig. 6), (iii) leads to increase in transport cost and (iv) indirectly increases the feedstock price. To reduce the hurdles faced in handling, transport and storage, baling or chopping of straw was the best method. The bulk density of loose straw, hammer milled and baled straw is 20–40, 20–100 and 110–200 kg/m^3, respectively (Sokhansanj and Hess 2009; Kargbo et al. 2010).

The most attractive and viable option for valorizing agro residues is biorefining. The paddy straw-based biorefinery scheme is presented in Fig. 7a–c. In comparision with biochemical the thermochemical route, it was found that the latter method leads to less productive in terms of biochemicals production. The biochemical route can be used as feedstock for production of multiple bioproducts such as bioethanol/butanol, biochemical and biocomposite materials or fuel. Among the thermochemical conversion process, the combustion and pyrolysis cannot be suitable for biorefinery approach if they used at the first stage of biorefinery conversion. The combustion of the plant wastes produces only heat and ashes. The yields of pyrolysis of straw can be in the forms of oil/solid/gas, which depends on reaction time, temperature and heating rate used in the process. Agglomeration, deposits formulation, fouling, corrosion issues and wear and tear of equipment are closely associated with the presence of mineral elements in lignocellulosic waste via thermochemical conversion method. To overcome these issues, the element should be extracted and reused (Dodson et al. 2013). Biomass pretreatment is used to break the lignin barrier and improve the enzymes accessibility of sugars. Biomass pretreatment of paddy straw is an inevitable process in the biochemical conversion route focused on bioalcohol (bioethanol or biobutanol) production. At the end of pretreatment process, hydrolysate, pretreated biomass and soluble lignin are produced. The pretreated straw contains higher cellu-

Fig. 6 Paddy straw transported from the agricultural field to the storage yard

lose content which should undergo enzymatic hydrolysis to yield fermentable sugars, and further, it was fermented to produce bioethanol. Hydrolysate can be used to produce chemicals with help of suitable microbial strain. The separated lignin can be used as feedstock for chemicals production or can be used as fuel due to its higher calorific value.

8 Conclusion

Agro residues are unavoidable by-products generated in the agricultural crop production. The annual agro residues generation is gradually increased to meet the food requirement of ever increasing world population. The major reasons for the farmers forced to adopt the onsite burning of agro residues are short period for land preparation for next crop cultivation, less straw price and higher transport charges. Valorization of single feedstock for single product production is not an economically viable project for agro residues. To reduce the environmental impacts of onsite stubble burning, biorefinery is a viable option to fractionating multiple products from these residues. Application of biorefinery approach for heterogeneous compositions of agro residues involves more challenges in selection process design and combining

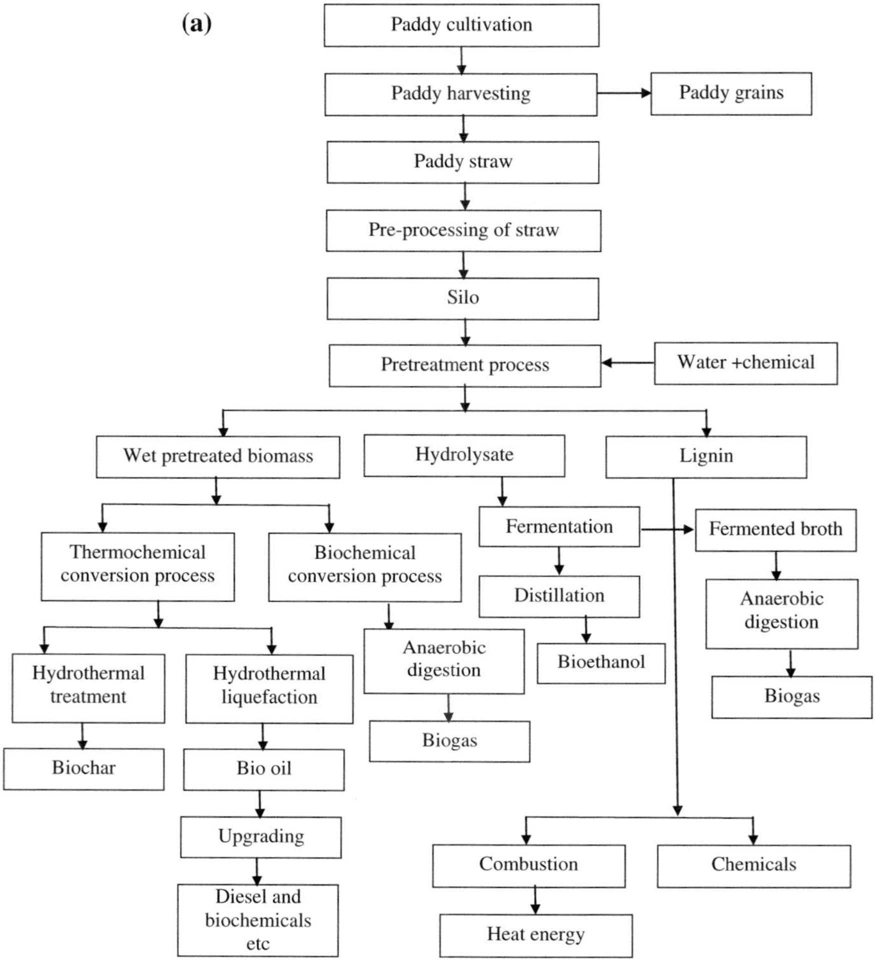

Fig. 7 **a** Biochemical conversion of straw into biofuel production. **b** Thermochemical conversion of straw into biofuel production. **c** Bio- and thermochemical conversion of straw into biofuel and biochemicals production

different biomass conversion technologies. The paddy straw is underutilized potential candidate to produce the bioethanol and value-added chemicals via biorefinery approach. Successful implementation of paddy straw-based biorefinery industries is depended not only on usage of technically economically viable technologies and but also on good supply chain management. Establishment of paddy straw biorefinery industries in India would lead to support the nation's energy security, sustainability and safe disposal of huge amount of waste.

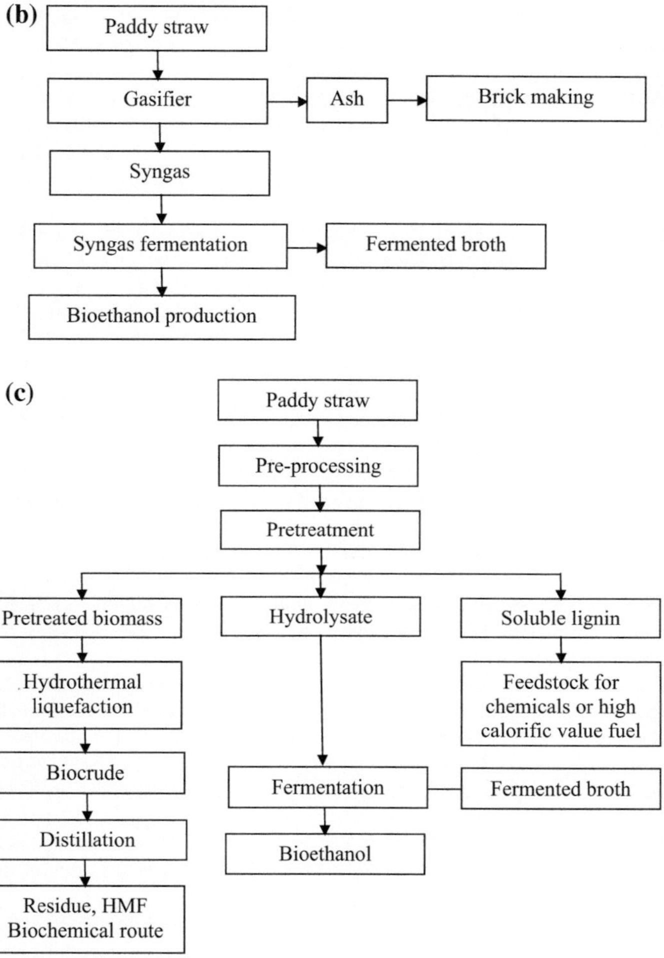

Fig. 7 (continued)

References

Adney WS, Rivard CJ, Ming SA, Himmel ME (1991) Anaerobic digestion of lignocellulosic biomass and wastes. Cellulases and related enzymes. Appl Biochem Biotechnol 30:165–183

Ahring BK, Westermann P (2007) Coproduction of bioethanol with other biofuels. Adv Biochem Eng Biotechnol 108:289–302. https://doi.org/10.1007/10_2007_067

Asadullah M, Rahman MA, Ali MM, Motin MA, Sultan MB, Alam MR et al (2008) Jute stick pyrolysis for bio-oil production in fluidized bed reactor. Bioresour Technol 99:44–50

Ballesteros I, Negro M, Oliva JM, Cabanas A, Manzanares P, Ballesteros M (2006) Ethanol production from steam explosion pretreated wheat straw. Appl Biochem Biotechnol 130:278–288

Beller HR, Lee TS, Katz L (2015) Natural products as biofuels and bio-based chemicals: fatty acids and isoprenoids. Nat Prod Rep 32:1508–1526. https://doi.org/10.1039/c5np00068h

Bertero M, de la Puente G, Sedran U (2012) Fuels from bio-oils: bio-oil production from different residual sources, characterization and thermal conditioning. Fuel 95:263–271

Bharathiraja B, Jayamuthunagai J, Sudharsanaa T, Bharghavi A, Praveenkumar R, Chakravarthy M, Yuvaraj D (2017) Biobutanol—an impending biofuel for future: a review on upstream and downstream processing techniques. Renew Sustain Energy Rev 68:788–807

De Bhowmick G, Sarmah AK, Sen R (2018) Lignocellulosic biorefinery as a model for sustainable development of biofuels and value added products. Bioresour Technol 247:1144–1154. https://doi.org/10.1016/j.biortech.2017.09.163

Bilal M, Asgher M, Iqbal HM, Hu H, Zhang X (2017a) Biotransformation of lignocellulosic

Bilal M, Asgher M, Iqbal HM, Hu H, Zhang X (2017b) Biotransformation of lignocellulosic materials into value-added products—a review. Int J BiolMacromol 98:447–458

Binod P, Sindhu R, Singhania RR, Vikram S, Devi L, Nagalakshmi S, Kurien N, Sukumaran RK, Pandey A (2010) Bioethanol production from rice straw: an overview. Bioresour Technol 101:4767–4774

Biomass Knowledge Portal. http://biomasspower.gov.in/. (as per May 20, 2016)

Bouaid A, Martinez M, Aracil J (2010) Biorefinery approach for coconut oil valorisation: a statistical study. Bioresour Technol 101:4006–4012

Braga RM, Melo DMA, Aquino FM, Freitas JCO, Melo MAF, Barros JMF, Fontes MSB (2013) Characterization and comparative study of pyrolysis kinetics of the rice husk and the elephant grass. J Therm Anal Calor. 115:1915–1920

Butler E, Devlin G, Meier D, McDonnell K (2013) Characterisation of spruce, salix, miscanthus & wheat straw for pyrolysis applications. Bioresour Technol 131:202–209

Byun J, Han J (2016) Catalytic production of biofuels (butene oligomers) and biochemicals (tetrahydrofurfuryl alcohol) from corn stover. Bioresour Technol 211:360–366. https://doi.org/10.1016/j.biortech.2016.03.123

Cabeza A, Piqueras CM, Sobron F, Garcia-Serna J (2016) Modeling of biomass fractionation in a lab-scale biorefinery: solubilization of hemicellulose and cellulose from holm oak wood using subcritical water. Bioresour Technol 200:90–102. https://doi.org/10.1016/j.biortech.2015.09.063

Caglar A, Demirbas A (2001) Conversion of cotton cocoon shell to liquid products by supercritical fluid extraction and low pressure pyrolysis in the presence of alkalis. Energy Convers Manage 42:1095–1104

Cai J, He Y, Yu X, Banksb Scott W, Yang Y, Zhang X, Yu Y, Liu R, Bridgwater Anthony V (2017) Review of physicochemical properties and analytical characterization of lignocellulosic biomass. Renew Sustain Energy Rev 76:309–322

Chen D, Cen K, Jing X, Gao J, Li C, Ma Z (2017) An approach for upgrading biomass and pyrolysis product quality using a combination of aqueous phase bio-oil washing and torrefaction pretreatment. Bioresour Technol 233:150–158. https://doi.org/10.1016/j.biortech.2017.02.120

Chen WH, Pen BL, Yuz CT, Hwang WS (2011a) Pretreatment efficiency and structural characterization of rice straw by an integrated process of dilute-acid and steam explosion for bioethanol production. Bioresour Technol 102:2916–2924

Chen WH, Pen BL, Yu CT, Hwang WS (2011b) Pretreatment efficiency and structural characterization of rice straw by an integrated process of dilute-acid and steam explosion for bioethanol production. Bioresour Technol 102:2916–2924

Chen G, Yao J, Yang H, Yan B, Chen H (2015) Steam gasification of acid-hydrolysis biomass CAHR for clean syngas production. Bioresour Technol 179:323–330. https://doi.org/10.1016/j.biortech.2014.12.039

Christopher M, Mathew AK, Kiran Kumar M, Pandey A, Sukumaran RK (2017) A biorefinery-based approach for the production of ethanol from enzymatically hydrolysed cotton stalks. Bioresour Technol . https://doi.org/10.1016/j.biortech.2017.03.190

de Jong E, Jungmeier G (2015) Biorefinery concepts in comparison to petrochemical refineries. In: Industrial biorefineries & white biotechnology, pp 3–33

Dhyani V, Bhaskar T (2017) A comprehensive review on the pyrolysis of lignocellulosic biomass. Renew Energy

Diaz AB, Moretti MMD, Bezerra-Bussoli C, Nunes CDCC, Blandino A, da Silva R et al (2015) Evaluation of microwave-assisted pretreatment of lignocellulosic biomass immersed in alkaline glycerol for fermentable sugars production. Bioresour Technol 185:316–323

Dodson JR, Li X, Zhou X, Zhao K, Sun N, Atahan P (2013) Origin and spread of wheat in China. Quatern Sci Rev 72:108–111

Dominguez E, Romani A, Alonso JL, Parajo JC, Yanez R (2014) A biorefinery approach based on fractionation with a cheap industrial by-product for getting value from an invasive woody species. Bioresour Technol 173:301–308. https://doi.org/10.1016/j.biortech.2014.09.104

ECN - Phyllis2 Database for biomass and waste (www.ecn.nl/phyllis2)

Esteghlalian A, Hashimoto AG, Fenske JJ, Penner MH (1997) Modeling and optimization of the dilute sulfuric acid pretreatment of corn stover, poplar and switchgrass. BioresourTechnol 59:129–136

Eynde EV, Lenaerts B, Tytgat T, Blust R, Lenaerts S (2016) Valorization of flue gas by combining photocatalytic gas pretreatment with microalgae production. Environ Sci Technol 50:2538–2545. https://doi.org/10.1021/acs.est.5b04824

Friedl A, Padouvas E, Rotter H, Varmuza K (2005) Prediction of heating values of biomass fuel from elemental composition. Anal Chim Acta 544:191–198

Gao H, Chen X, Wei J, Zhang Y, Zhang L, Chang J et al (2016) Decomposition dynamics and changes in chemical composition of wheat straw residue under anaerobic and aerobic conditions. PLoS ONE 11(7):e0158172

Garcia-Cubero MT, González-Benito G, Indacoechea I, Coca M, Bolado S (2009) Effect of ozonolysis pretreatment on enzymatic digestibility of wheat and rye straw. Bioresour Technol 100:1608–1613

Gaurava N, Sivasankarib S, Kiranc GS, Ninawea A, Selvinb J (2017) Utilization of bioresources for sustainable biofuels: a review. Renew Sustain Energy Rev 73:205–214

Hellier P, Purton S, Ladommatos N (2015) Molecular structure of photosynthetic microbial biofuels for improved engine combustion and emissions characteristics. Front Bioeng Biotechnol 3:49. https://doi.org/10.3389/fbioe.2015.00049

Henry RJ (2018) Biofuels from crop plants. Encyclopedia of applied plant sciences, 2nd edn, vol 3, pp 177–179. http://dx.doi.org/10.1016/B978-0-12-394807-6.00169-6

Hiloidhari M, Baruah D (2011a) Crop residue biomass for decentralized electrical power generation in rural areas (part 1): investigation of spatial availability. Renew Sustain Energy Rev 15:1885–1892

Hiloidhari M, Baruah DC (2011b) Rice straw residue biomass potential for decentralized electricity generation: a GIS based study in Lakhimpur district of Assam, India. Energy Sustain Dev 15:214–222

Hiloidhari M, Das D, Baruah DC (2014) Bioenergy potential from crop residue biomass in India. Renew Sustain Energy Rev 32:504–512

IARI (2012) Crop residues management with conservation agriculture: potential, constraints and policy needs. Indian Agricultural Research Institute, New Delhi, vii + 32 p

Jekayinfa SO, Scholz V (2009) Potential availability of energetically usable crop residues in Nigeria. Energy Sources Part A Recovery Utilization Environ Eff 31:687–697

Kargbo F, Xing J, Zhang Y (2010) Property analysis and pretreatment of rice straw for energy use in grain drying: a review. Agric Biol J North Am 1(3):195–200

Kis D, Sucic B, Guberac V, Voca N, Rozman V, Sumanovac L (2009) Soybean biomass as a renewable energy resource. Agric Conspec Sci 74:201–203

Laghari SM, Isa MH, Laghari AJ (2016) Delignification of palm fiber by microwave assisted chemical pretreatment for improving energy efficiency. Malays J Sci 35:8–14

Le DM, Sørensen HR, Knudsen NO, Schjoerring JK, Meyer AS (2014) Biorefining of wheat straw: accounting for the distribution of mineral elements in pretreated biomass by an extended pretreatment-severity equation. Biotechnol Biofuels 7(1):141

Ludueña L, Fasce D, Alvarez VA, Stefani PM (2011) Nanocellulose from rice husk following alkaline treatment to removesilica. BioResources 6:1440–1453

Mani S, Tabil LG, Sokhansanj S (2006a) Effects of compressive force, particle size and moisture content on mechanical properties of biomass pellets from grasses. Biomass Bioenergy 30:648–654
Mani S, Tabil LG, Sokhansanj S (2006b) Effects of compressive force, particle size and moisture content on mechanical properties of biomass pellets from grasses. Biomass Bioenergy 30:648–654
McKendry P (2002) Energy production from biomass (part 1): overview of biomass. Bioresour Technol 83(1):37–46
Minowa T, Kondo T, Sudirjo ST (1998) Thermochemical liquefaction of Indonesian biomass residues. Biomass Bioenergy 14:517–524
Motte JC, Escudie R, Beaufils N, Steyer JP, Bernet N, Delgenes JP, Dumas C (2014) Morphological structures of wheat straw strongly impacts its anaerobic digestion. Ind Crops Prod 52:695–701
National Renewable Energy Laboratory (2009) Renewable energy data book: energy efficiency & renewable energy
Nigam PS, Gupta N, Anthwal A (2009) Pre-treatment of agro-industrial residues. In: Nigam PS, Pandey A (eds) Biotechnology for agro-industrial residues utilization, 1 edn. Springer, Netherlands, pp 13–33
Nordin R, Said CMS, Ismail H (2007) Properties of rice husk powder/natural rubber composite. Solid State Sci Technol 15:83–91
Octave S, Thomas D (2009) Biorefinery: toward an industrial metabolism. Biochimie 91:659–664
Phan BMQ, Duong LT, Nguyen VD, Tran TB, Nguyen MHH, Nguyen LH, Nguyen DA, Luu LC (2014) Evaluation of the production potential of bio-oil from Vietnamese biomass resources by fast pyrolysis. Biomass Bioenergy 62:74–81
Pilon G (2007) Utilization of Arecanut (*Areca catechu*) husk for gasification. [MSc. thesis]. Department of Bioresource Engineering, McGill University, Montreal
Prassad S, Singh A, Joshi HC (2007a) Ethanol as an alternative fuel from agricultural, industrial and urban residues. Resourc Conserv Recycl 50:1–39
Prassad S, Singh A, Joshi HC (2007b) Ethanol as an alternative fuel from agricultural, industrial and urban residues. Resourc Conserv Recycl 50:1–39
Rahman MT (2006) Green energy development model in the St.Martin's Island and energy from coconut palm biomass, ISBN: 984-8323-02-3
Rai SN, Walli TK, Gupta BN (1989) The chemical composition and nutritive value of rice straw after treatment with urea or *Coprinus fimetarius* in a solidstate fermentation system. Anita Feed Sci Technol 26:81–92
Raman S, Mohr A, Helliwell R, Ribeiro B, Shortall O, Smith R, Millar K (2015) Integrating social and value dimensions into sustainability assessment of lignocellulosic biofuels. Biomass Bioenergy 82:49–62. https://doi.org/10.1016/j.biombioe.2015.04.022
Raveendran K, Ganesh A, Khilar C (1995) Influence of mineral matter on biomass pyrolysis characteristics. Fuel 74:1812–1822
Saini JK, Saini R, Tewari L (2015) Lignocellulosic agriculture wastes as biomass feedstocks for second-generation bioethanol production: concepts and recent developments. Biotech 5:337–353
Sanchez C (2009) Lignocellulosic residues: biodegradation and bioconversion by fungi. Biotechnol Adv 27:185–194
Sarnklong C, Cone JW, Pellikaan W, Hendriks WH (2010) Utilization of rice straw and different treatments to improve its feed value for ruminants: a review. Asian Aust J Anim Sci 23(5):680–692
Singh G, Ahuja N, Batish M, Capalash N, Sharma P (2008a) Biobleaching of wheat straw-rich soda pulp with alkalophilic laccase from gamma-proteobacterium JB: optimization of process parameters using response surface methodology. Bioresour Technol 99:7472–7479. https://doi.org/10.1016/j.biortech.2008.02.023
Singh J, Panesar BS, Sharma SK (2008b) Energy potential through crop biomass using geographical information system—a case study of Punjab. Biomass Bioenergy 32:301–307
Singh RP, Dhaliwal HS, Sidhu HS, Manpreet-Singh YS, Blackwell J (2008c) Economic assessment of the Happy Seeder for rice-wheat systems in Punjab, India. In: Conference Paper, AARES 52nd Annual conference. ACT, Canberra. Australia

Sokhansanj S, Hess JR (2009) Biomass supply logistics and infrastructure. In: Biofuels, Humana Press, Totowa, NJ, pp 1–25

State of Indian Agriculture (2015–2016) Ministry of Agriculture & Farmers Welfare, Government of India

Sun Y, Cheng JJ (2005) Dilute acid pretreatment of rye straw and Bermuda grass for ethanol production. Bioresour Technol 96:1599–1606

Tao L, Aden A, Elander RT (2013) Economics of pretreatment for biological processing. In: Aqueous pretreatment of plant biomass for biological and chemical conversion to fuels and chemicals. Wiley, Hoboken, pp 311–333. https://doi.org/10.1002/9780470975831.ch15

Thomas P, Soren N, Rumjit NP, James JG, Saravanakumar MP (2017) Biomass resources and potential of anaerobic digestion in Indian scenario. Renew Sustain Energy Rev 77:718–730

WBA Global Bioenergy Statistics (2017). www.worldbioenergy.org. (accessed on 31.1.2018)

Wang MJ, Huang YF, Chiueh PT, Kuan WH, Lo SL (2012) Microwave-induced torrefaction of rice husk and sugarcane residues. Energy 37:177–184

Wilaipon P (2009) The effects of briquetting pressure on banana-peel briquette and the banana waste in Northern Thailand. Am J App Sci 6:167–171

World Energy Outlook (2002) International Energy Agency, IEA/OECD, Paris

Xu J, Xiong P, He B (2016) Advances in improving the performance of cellulase in ionic liquids for lignocellulose biorefinery. Bioresour Technol 200:961–970. https://doi.org/10.1016/j.biortech.2015.10.031

Zabaniotou A, Ioannidou O, Antonakou E, Lappas A (2008) Experimental study of pyrolysis for potential energy, hydrogen and carbon material production from lignocellulosic biomass. Int J Hydrogen Energy 33:2433–2444

Zhang M, Xie L, Yin Z, Khanal SK, Zhou Q (2016) Biorefinery approach for cassava-based industrial wastes: current status and opportunities. Bioresour Technol 215:50–62. https://doi.org/10.1016/j.biortech.2016.04.026

Green Nanotechnology for Biofuel Production

Susana Rodríguez-Couto

Abstract The current usage of fossil fuel is not sustainable and, in addition to this, has well-known negative impacts for the environment such as the greenhouse effect and the depletion of the ozone layer. Therefore, the search for alternative energy sources to meet the ever-increasing world's energy demand is a top priority, especially in the most industrialized countries. In this sense, lignocellulosic biomass, not competing with food production and being available at very low or null cost, is an attractive and economic alternative source for the production of biofuels. Additionally, the new emerging nanotechnologies are increasingly being applied to convert biomass into biofuels. In this chapter, the application of the new emerging green nanotechnologies to biofuel production is pointed out.

Keywords Biocatalyst · Biofuels · Lignocellulosic biomass · Nanotechnology Renewable energy

1 Introduction

Nowadays, the main energy sources for human activity come from fossil and mineral fuels, nuclear and hydroelectric sources. Renewable energy represents only about 10%, thus the world remains highly dependent on fossil fuels (Fig. 1) (World Energy Resources 2013). However, fossil-based energies are very harmful to the environment causing global warming, ozone layer depletion, biosphere and geosphere destruction, and ecological devastation. Thus, the current energy production can be considered as a noxious industry in terms of both pollution generation and environmental impact

S. Rodríguez-Couto (✉)
Ceit-IK4, Paseo Manuel de Lardizábal 15, 20018 San Sebastian, Spain
e-mail: srodriguez@ceit.es

S. Rodríguez-Couto
Universidad de Navarra, Tecnun, Paseo Manuel de Lardizábal 13, 20018 San Sebastian, Spain

S. Rodríguez-Couto
KERBASQUE, Basque Foundation for Science, Maria Diaz de Haro 3, 48013 Bilbao, Spain

N. Srivastava et al. (eds.), *Sustainable Approaches for Biofuels Production Technologies*, Biofuel and Biorefinery Technologies 7,
https://doi.org/10.1007/978-3-319-94797-6_4

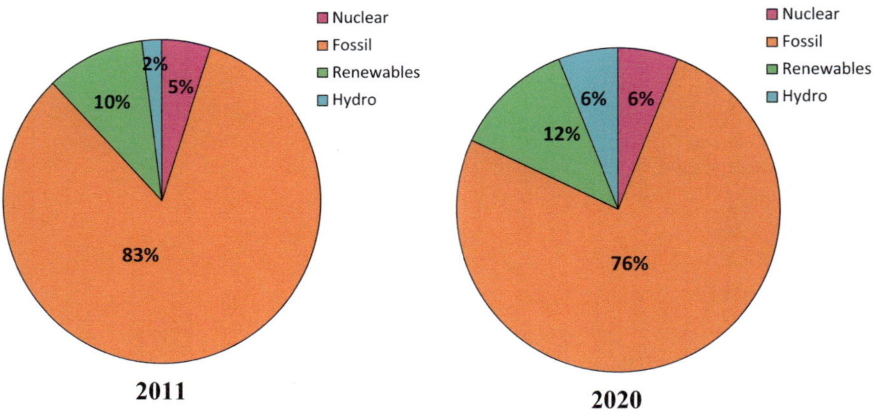

Fig. 1 Total primary energy supply by resource 2011 and 2020 (World Energy Resources 2013)

since the industrial revolution in the eighteenth century (Serrano et al. 2009). Hence, about 80% of the CO_2 emissions worldwide are generated by the energy sector. Thus, fossil fuels are responsible for around 27 billion ton of CO_2 emissions every year. This has impelled the search for alternative energy sources to replace fossil fuels (Rutz and Janssen 2007; Demirbas 2008). Among the primary alternative sources that can replace fossil fuels, biomass has appeared as an appalling alternative in the last years due to its availability, low or null cost, renewability, and nontoxicity (Huang et al. 2015). Also, lignocellulosic biomass is generated in large amount from various industries as an undesired waste with zero market value, so there is no feedstock scarcity in comparison to fossil fuels. However, the technologies for biomass processing present technological and economical limitations (Masran et al. 2016). This has driven the interest in the emerging nanotechnologies to improve the existing technologies as well as to provide new alternatives to fulfill the world's energy needs. The term nanotechnology designs materials and phenomena occurring at nanoscale (anything measuring between 1 and 100 nm). The properties of nanomaterials/nanosystems are very different from those of the bulk materials. Thus, the nanoworld is mainly controlled by quantum mechanics whose rules are very different from classical physics. This means that the behavior of materials at nanoscale will likely be very different from their bulk counterparts and present particular properties (Nanotechnology 2014).

Nanotechnology can be applied to feedstock modification and the development of more efficient catalysts (Antunes et al. 2017). In addition, immobilization of enzymes on nanomaterials (e.g., lipase, cellulase) can make feasible the conversion of lignocellulosic feedstock into biodiesel and bioethanol. Thus, Guo et al. (2012) proposed the use of solid acid-functionalized paramagnetic nanoparticles to convert cellulosic materials into soluble sugars, which are platform molecules for biofuel generation, as indicated in Fig. 2. Nanocatalysts combined the positive characteristics of both homogeneous and heterogeneous catalysts (Fig. 3) (Lee and Juan 2017). Thus,

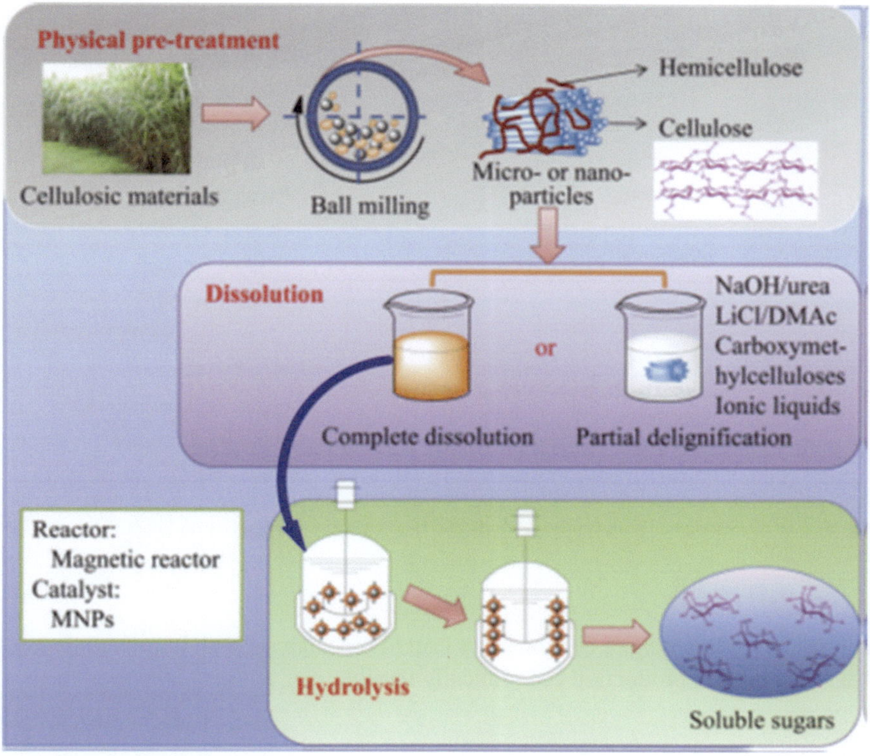

Fig. 2 Hydrolysis of cellulosic material into soluble sugars by metal nanoparticles. Reprinted from Guo et al. (2012), with kind permission from Elsevier Ltd., UK

nanocatalysts allow rapid, selective, and highly active reactions together with easy catalyst recycling (Singh and Tandon 2014). In addition, their high volume/size ratio will make it possible to load a great amount of enzyme per weight of support (Gupta et al. 2011). Consequently, they are having great attention in the field of enzyme immobilization.

A nanotechnology process to immobilize the costly enzymes used to convert cellulose into sugars was developed in 2009 at the Louisiana Tech University (Ruston, Louisiana, USA). This technology allows reusing the enzymes several times, thereby, reducing significantly the overall cost of the process. More recently (Bacik et al. 2017), a multi-research team composed of researchers from Los Alamos National Laboratory (Los Alamos, New Mexico, USA), Oak Ridge National Laboratory (Oak Ridge, Tennessee, USA), Lawrence Berkeley Laboratory (Berkeley, California, USA), and the Norwegian University of Life Sciences (Oslo, Norwegian) has mapped the three-dimensional structure of a copper-dependent enzyme called lytic polysaccharide monooxygenase (LPMO). This enzyme breaks down polysaccharides, such as those of the lignocellulose biomass. Therefore, understanding the

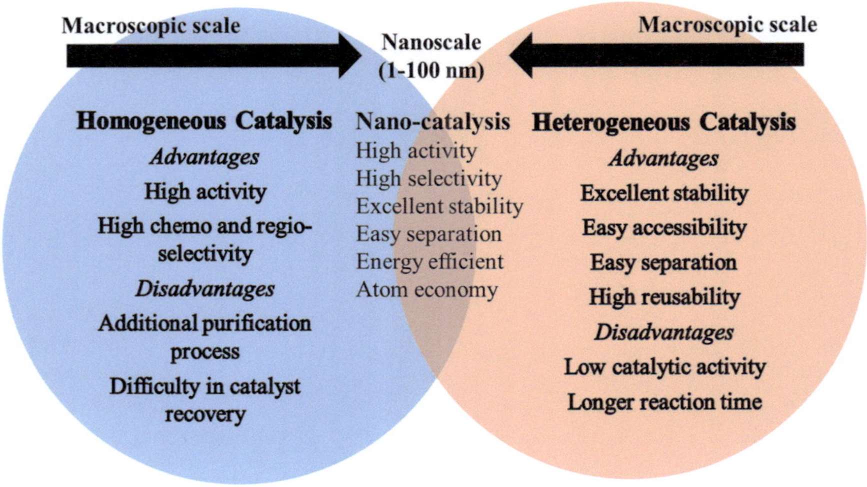

Fig. 3 Characteristics of bulk (homogeneous and heterogeneous) catalysts and nanocatalysts (Lee and Juan 2017)

mechanism of this type of enzymes can lead to biocatalysts with improved features that make biofuel production economically feasible.

2 Green Nanotechnologies

The promising use of nanomaterials is hampered by the harmful effects they may pose to human health and environment (Alvarez 2006; Maynard et al. 2006). Consequently, it is of utmost importance to know whether nanomaterials or their fabrication processes have negative or undesirable impacts that make their use impractical. In this regard, the principles of green chemistry and green engineering have to be used utilizing a life cycle approach that considers the total range of economic, environmental, and societal implications. Thus, to develop nanotechnology in a manner that provides benefits for society avoiding harm to health and environment, some researchers have applied the principles of green chemistry to nanoscience (Dahl et al. 2007; Morose 2010; Gilberston et al. 2015). Hence, the aim of green nanotechnology is to reduce or avoid the potential undesirable impacts of nanomaterials or their production processes. For this, the product lifecycle (i.e., from extraction to end-of-life) should be considered (Fig. 4) (Dahl et al. 2007; Gilberston et al. 2015; Hutchison 2016).

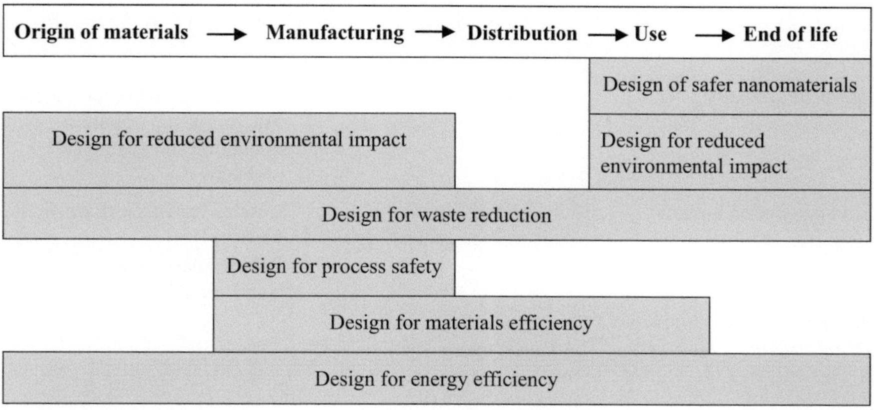

Fig. 4 Mapping of the green nanoscience/nanotechnology design principles across the stages of the lifecycle of products (Hutchison 2016)

3 Current Status of Biofuel Production

Currently, biofuels (i.e., bioethanol and biodiesel) have been manufactured in different countries, Unites States of America (USA), Brazil and the European Union (EU) being the largest producers (Statista 2016; Manochio et al. 2017). These first-generation biofuels are produced from different food feedstocks (Naik et al. 2010). Thus, in USA ethanol is produced from corn, whereas Brazil produces it from sugarcane and the EU from sugar beet (Manochio et al. 2017). As for biodiesel, in the EU, it is mainly produced from canola and sunflower oils. However, increased cost of food and feedstuffs has promoted the second-generation biofuels from nonfood feedstock (Patumsawad 2011; Eggert and Greaker 2014). In this sense, lignocellulosic biomass is very promising since it is abundant worldwide, no extra land is needed and does not interfere with food and feed production. Thus, the global production of plant biomass is about 200×10^9 tons per year, of which around 8×10^9–20×10^9 tons per year can be utilized for biofuel production (Kuhad and Singh 1993; Saini et al. 2014). Although second-generation biofuels have several advantages (Table 1) (Naik et al. 2010), there are still some issues regarding cost and technology that remain to be solved (Fig. 5).

Significant advances have been attained in the past years in all aspects of lignocellulose conversion into ethanol. Companies such as Betarenewables/Biochemtex (Crescentino, Italy), Inbicon/Dong Energy (Kalundborg, Denmark), Abengoa (Babilafuente, Spain), Clariant (Straubing, Germany) in Europe, Abengoa (Hugoton, Kansas), Dupont (Nevada, Iowa) and POET-DSM (Emmetsburg, Iowa) in USA, Iogen Corporation in Canada and GranBio (Alagoas), and Raízen (Piracicaba) in Brazil has started to commercialize cellulosic ethanol. The existent commercial, demonstration, and pilot plants are essential industrial platforms to overcome the

Table 1 Comparison of petroleum fuel, first and second-generation biofuel (Naik et al. 2010)

	Advantages	Hurdles
Petroleum refinery		Depletion of petroleum reserve Environmental pollution Economic and ecological problems
First-generation biofuels	Environmentally friendly Economic and social security	Limited feedstock (food vs. fuel) Blended partly with conventional fuel
Second-generation biofuels	Not competing with food Advance technology still under development to reduce the cost of conversion Environmentally friendly	

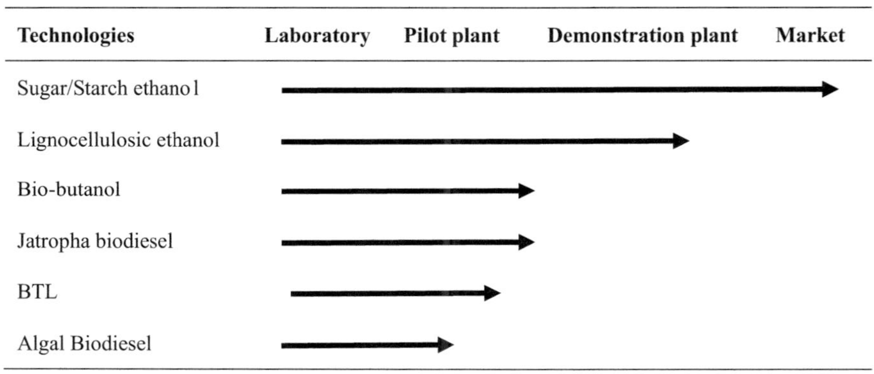

Fig. 5 World biofuel technology status (Joshi et al. 2017)

bottlenecks and barriers to the full commercialization of the second-generation bioethanol in the near future.

The *World Energy Outlook 2009* (IEA 2009) *450 Scenario1* predicts biofuels will supply 9% (11.7 EJ) of the total transport fuel demand (126 EJ) in 2030. The *Blue Map Scenario*2 of *Energy Technology Perspectives 2008* (IEA 2008), which extends the analysis up to 2050, predicts biofuels will supply 26% (29 EJ) of the total transportation fuel demand (112 EJ) in 2050, with second-generation biofuels standing for approximately 90% of all biofuel. More than half of the second-generation biofuel production in the *Blue Map Scenario* is thought to take place in non-OECD (Organization for Economic Cooperation and Development) countries, 19% of the total production being attributed to China and India.

Table 2 Biofuels produced using green nanotechnology

Feedstock	Biocatalyst	Product	Reference
Spent tea (*Camellia sinensis*)	Cobalt nanoparticles	Biodiesel (40.8%) Bioethanol (57.5%)	Mahmood and Hussain (2010)
Microcrystalline cellulose	Cellulase physisorbed on silica nanoparticles	Bioethanol	Lupoi and Smith (2011)
Jackfruit waste and sugarcane leaves	Cellulase immobilized on MnO_2 nanoparticles	Bioethanol (21.96 g/L)	Cherian et al. (2015)
Sesbania aculeate	Cellulase bound magnetic nanoparticles	Bioethanol (5.31 g/L)	Baskar et al. (2016)

4 Biofuel Production with Green Nanotechnologies

The key bottleneck for lignocellulosic-derived biofuels is the lack of technology to convert biomass into liquid fuels efficiently. Therefore, the development of efficient technologies to solve this problem is an urgent need. In this regard, advances in nanotechnology, which have made possible to understand and control chemistry at the molecular scale, augur the potential development of efficient biomass-to-fuels production technologies. Thus, several nanomaterials such as TiO_2, Fe_3O_4, SnO_2, ZnO, carbon, graphene, and fullerene have been applied to biofuel production due to their unique properties. In addition, immobilization of different enzymes, such as cellulases and hemicellulases, on several nanomaterials has been applied to bioethanol production (Antunes et al. 2017 and references therein). This strategy allows recycling the enzymes, thus reducing the process cost. However, the release of nanoparticles into the environment supposes a serious threat to human health and environment (Gupta et al. 2015). Therefore, a life cycle approach in order to detect and mitigate the environmental impacts that may arise in nanomaterial production must be performed (Nahr et al. 2015).

As green nanotechnology is still in its initial stage, there are few studies on the production of biofuels from lignocellulose biomass applying such a technology. In Table 2, recent studies on the production of biofuels from lignocellulosic biomass using green nanotechnology are presented. Thus, Mahmood and Hussain (2010) reported the conversion of spent tea (solid waste) into biofuels by using nanocatalytic gasification followed by transesterification of the liquid fraction to obtain biodiesel and *Aspergillus niger* fermentation of the solid fraction to produce bioethanol.

Lupoi and Smith (2011) reported higher ethanol yields in a process consisting of simultaneous saccharification and solid-state fermentation (SSF) reactions of microcrystalline cellulose when cellulase was physisorbed on silica nanoparticles than those attained with the free enzyme. Therefore, their results showed that the use of silica-immobilized cellulase increased ethanol yields in the conversion of lignocellulosic materials by SSF.

Cherian et al. (2015) studied the hydrolysis and bioethanol production efficiency of different organic wastes by cellulase from *Aspergillus fumigatus* immobilized on MnO_2 nanoparticles in combination with yeast for SSF. They found that the immobilized enzyme led to a higher bioethanol yield (22 g/L) than that of the free enzyme (18 g/L). In addition, immobilization widened cellulase operation range of pH and temperature and made cellulase reutilization possible.

Baskar et al. (2016) reported the efficient production of bioethanol by the hydrolysis of *Sesbania aculeate* biomass using cellulose-bound magnetic nanoparticles. Thus, a maximum bioethanol production of 5.3 g/L was attained operating under optimal conditions (i.e., a nanobiocatalyst concentration of 1.5% (w/v), a biomass concentration of 4% (w/v), and a temperature of 30 °C).

Biofuels derived from microalgae (third-biofuel generation) also hold great potential. Algae biomass is formed up to 50% oil, making them very suitable for biodiesel production. In addition, the algae carbohydrates can be used for bioethanol production. However, several challenges such as water requirements of algae to grow, the low concentration of natural CO_2 in the atmosphere and the high cost of microalgae cultivation and harvesting (Kim et al. 2013; Rashid et al. 2014) must be overcome before converting algae into biofuels. The use of nanotechnology can help to solve some of these problems. Thus, nanomaterials have been used on lipid accumulation, extraction and on the transesterification process as a catalyst support or a catalyst (Andrijanto 2009; Lin 2009a, b; Lu 2009) and in cultivation and harvesting (Lee et al. 2015 and references therein). Thus, recently Duraiarasan et al. (2016) used enzymes (i.e., cellulose and lipase) immobilized on magnetic nanoparticles for biodiesel production from the marine microalga *Chlorella salina*. A maximum yield of 93.56% was produced under optimized conditions (i.e., 2 g of each immobilized enzyme for the whole process and 60% water content). The immobilized enzymes operated effectively for 10 successive cycles. Hence, the process holds potential for large-scale production of biodiesel from wet algal biomass in an easier, effective, and eco-friendly manner.

Despite the research efforts to apply nanoparticles to the production of biofuels from microalgae, there are several critical points, such as rational designs for high-efficiency and stable nanoparticles, cost-effective recycling and environmental concerns, that remain to be solved.

5 Concluding Remarks and Future Perspectives

With no doubt, it is necessary to find new alternative safe energy sources to replace fossil-based ones. Lignocellulosic biomass appears to be a very promising alternative due to its availability at low or null cost. This bioenergy is renewable and environmentally friendly but its conversion is still an arduous process. The use of nanotechnology appears as an interesting option to ease this process, paving the way for an efficient, environmentally friendly, and economically viable production of biofuels. However, before applying this technology, there are safety concerns regarding human health and environment that have to be thoroughly addressed.

References

Alvarez PJJ (2006) Nanotechnology in the environment-the good, the bad and the ugly. ASCE J Environ Eng 132:1233

Andrijanto E (2009) Novel bifunctional mesoporous silica nanomaterial catalyst for biodiesel production. Poster of University of Huddersfield Research Festival

Antunes FAF, Gaikwad S, Ingle AP, Pandit R, dos Santos JC, Rai M, da Silva SS (2017) Bioenergy and biofuels: nanotechnological solutions for sustainable production. In: Rai M, da Silva SS (eds) Nanotechnolgy for bioenergy and biofuel production. Springer, Switzerland, pp 3–19

Bacik JP, Mekasha S, Forsberg Z, Kovalevsky AY, Vaaje-Kolstad G, Eijsink VGH, Nix JC, Coates L, Cuneo MJ, Unkefer CJ, Chen JCH (2017) Neutron and atomic resolution X-ray structures of a lytic polysaccharide monooxygenase reveal copper-mediated dioxygen binding and evidence for N-terminal deprotonation. Biochemistry 56:2529–2532

Baskar G, Kumar RN, Melvin XH, Aiswarya R, Soumya S (2016) *Sesbania aculeate* biomass hydrolysis using magnetic nanobiocomposite of cellulase for bioethanol production. Renew Energ 98:23–28

Cherian E, Dharmendirakumar M, Baskar G (2015) Immobilization of cellulase onto MnO_2 nanoparticles for bioethanol production by enhanced hydrolysis of agricultural waste. Chin J Catal 36:1223–1229

Dahl JA, Maddux BLS, Hutchison JE (2007) Toward greener nanosynthesis. Chem Rev 107:2228–2269

Demirbas A (2008) Biofuels sources, biofuel policy, biofuel economy and global biofuel projections. Energ Convers Manag 49:2106–2116

Duraiarasan S, Razack SA, Manickam A, Munusamy A, Syed MB, Ali MY, Ahmed GM, Mohiuddin MdS (2016) Direct conversion of lipids from marine microalga *C. salina* to biodiesel with immobilised enzymes using magnetic nanoparticle. J Environ Chem Eng 4:1393–1398

Eggert H, Greaker M (2014) Promoting second generation biofuels: does the first generation pave the road? Energies 7:4430–4445

Gilbertson LM, Zimmerman JB, Plata DL, Hutchison JE, Anastas PT (2015) Designing nanomaterials to maximize performance and minimize undesirable implications guided by the principles of green chemistry. Chem Soc Rev 44:5758–5777

Guo F, Fang Z, Xu CC, Smith RL Jr (2012) Solid acid mediated hydrolysis of biomass for producing biofuels. Prog Energ Combust Sci 38:672–690

Gupta MN, Kaloti M, Kapoor M, Solanki K (2011) Nanomaterials as matrices for enzyme immobilization. Artif Cells Blood Substit Immobil Biotechnol 39:98–109

Gupta IR, Anderson AJ, Rai M (2015) Toxicity of fungal-generated silver nanoparticles to soil inhabiting *Pseudomonas putida* KT2440, a rhizospheric bacterium responsible for plant protection and bioremediation. J Hazard Mater 286:48–54

Huang L, Luo F, Chen Z, Megharaj M, Naidu R (2015) Green synthesized conditions impacting on the reactivity of Fe NPs for the degradation of malachite green. Spectrochim Acta A 137:154–159

Hutchison JE (2016) The road to sustainable nanotechnology: challenges, progress and opportunities. ACS Sustain Chem Eng 4:5907–5914

IEA (2008) Energy technology perspectives 2008: scenarios and strategies to 2050. OECD/IEA, France

IEA (2009) World energy outlook 2009. OECD/IEA, France

Joshi G, Pandey JK, Rana S, Rawat DS (2017) Challenges and opportunities for the application of biofuel. Renew Sust Energ Rev 79:850–866

Kim J, Yoo G, Lee H, Lim J, Kim K, Kim CW, Park MS, Yang JW (2013) Methods of downstream processing for the production of biodiesel from microalgae. Biotechnol Adv 31:862–876
Kuhad RC, Singh A (1993) Lignocellulose biotechnology: current and future prospects. Crit Rev Biotechnol 13:151–172
Lee HV, Juan JC (2017) Nanocatalysis for the conversion of nonedible biomass to biogasoline via deoxygenation reaction. In: Rai M, da Silva SS (eds) Nanotechnolgy for bioenergy and biofuel production. Springer, Switzerland, pp 301–323
Lee YC, Lee K, Oh YK (2015) Recent nanoparticle engineering advances in microalgal cultivation and harvesting processes of biodiesel production: a review. Bioresour Technol 184:63–72
Lin V (2009a) Nanofarming technology extracts biofuel oil without harming algae. News released from Office of Public Affairs
Lin V (2009b) Nanoparticle technology for biorefinery of non-food source feedstocks. http://www1.eere.energy.gov/industry/nanomanufacturing/pdfs/nanoparticle_tech_biorefinery.pdf
Lu F (2009) Developing innovative photobioreactor and extraction technology for production of biodiesel feedstock using microalgae. http://www.reeis.usda.gov/web/crisprojectpages/217877.html
Lupoi JS, Smith EA (2011) Evaluation of nanoparticle-immobilized cellulase for improved ethanol yield in simultaneous saccharification and fermentation reactions. Biotechnol Bioeng 108:2835–2843
Mahmood T, Hussain ST (2010) Nanobiotechnology for the production of biofuels from spent tea. Afr J Biotechnol 9:858–886
Manochio C, Andrade BR, Rodriguez RP, Moraes BS (2017) Ethanol from biomass: a comparative overview. Renew Sustain Energ Rev 80:743–755
Masran R, Zanirun Z, Bahrin EKB, Ibrahim MF, Yee PL, Abd-Aziz S (2016) Harnessing the potential of ligninolytic enzymes for lignocellulosic biomass pretreatment. Appl Microbiol Biotechnol 100:5231–5246
Maynard AD, Aitken RJ, Butz T, Colvin V, Donaldson K, Oberdorster G, Philbert MA, Ryan J, Seaton A, Stone V, Tinkle SS, Tran L, Walker NJ, Warheit DB (2006) Safe handling of nanotechnology. Nature 444:267–269
Morose G (2010) The 5 principles of design for safer nanotechnology. J Cleaner Prod 18:285–289
Nahr FK, Mokarram RR, Hejazi MA, Ghanbarzadeh BK, Mahmoud S, Benis KZ (2015) Optimization of the nanocellulose based cryoprotective medium toenhance the viability of freeze dried *Lactobacillus plantarum* using responsesurface methodology. LWT Food Sci Technol 64:326–332
Naik SN, Goud VV, Rout PK, Dalai AK (2010) Production of first and second generation biofuels: a comprehensive review. Renew Sustain Energ Rev 14:578–597
Nanotechnology (2014) http://science.howstuffworks.com/nanotechnology1.htm
Patumsawad S (2011) 2nd generation biofuels: technical challenge and R and D opportunity in Thailand. J Sustain Energ Environ (Special Issue):47–50
Rashid N, Rehman MSU, Sadiq M, Mahmood T, Han JI (2014) Current status, issues and developments in microalgae derived biodiesel production. Renew Sustain Energ Rev 40:760–778
Rutz D, Janssen R (2007) Biofuel technology handbook, pp 106–129
Saini JK, Saini R, Tewari L (2014) Lignocellulosic agriculture wastes as biomass feedstocks for second-generation bioethanol production: concepts and recent developments. 3 Biotech 4:337–353
Serrano E, Rus G, Garcia-Martinez J (2009) Nanotechnology for sustainable energy. Renew Sustain Energ Rev 13:2373–2384
Singh SB, Tandon PK (2014) Catalysis: a brief review on nano-catalyst. J Energy Chem Eng 2:106–115
Statista, The statistic portal (2016) The world's biggest biodiesel producers in 2014, by country (in billion liters), http://www.statista.com/statistics/271472/biodiesel-production-in-selectedcountries/
World Energy Resources (2013) Survey. World Energy Council, London

Biofuels from Protein-Rich Lignocellulosic Biomass: New Approach

Neha Srivastava, Manish Srivastava, S. N. Upadhyay, P. K. Mishra and Pramod W. Ramteke

Abstract Increasing consumption of fossil fuels and concern over environmental emissions has provided impetus to the development of renewable biofuels. Presently available biofuel production processes and developing approaches have focused on closing the carbon cycle by biological fixation of atmospheric carbon dioxide and conversion of biomass into biofuels. Lignocellulosic plant residues are found in abundance quantity and contain an appropriate amount of protein which is a by-product of biomass pretreatment. Besides conversion of carbohydrates into fuel, efforts towards conversion protein to fuel and ammonia may improve its value addition. Moreover, development of this technology will also realize its advantages of high carbon fixation rates, reduce consumption of synthetic fertilizer, inexpensive, and simple feedstock processing. Therefore, the present chapter provides an overview of the production process of biofuels using lignocellulosic plant residues and their protein by-products. The major hurdles to enhance the yield/production of the product and possible approaches to overcome these hindrances were also discussed.

Keywords Biofuels · Protein waste · Nitrogen fertilizer · Ammonia recycling
Protein conversion

N. Srivastava (✉) · S. N. Upadhyay · P. K. Mishra (✉)
Department of Chemical Engineering and Technology, Indian Institute of Technology, (Banaras Hindu University), Varanasi 221005, India
e-mail: sri.neha10may@gmail.com

P. K. Mishra
e-mail: pkmishra.che@itbhu.ac.in

M. Srivastava
Department of Physics and Astrophysics, University of Delhi, Delhi 110007, India

P. W. Ramteke
Department of Biological Sciences, Sam Higginbottom University of Agriculture Technology and Sciences (Formerly Allahabad Agricultural Institute) Deemed to be University, Allahabad, Uttar Pradesh 221007, India

N. Srivastava et al. (eds.), *Sustainable Approaches for Biofuels Production Technologies*, Biofuel and Biorefinery Technologies 7,
https://doi.org/10.1007/978-3-319-94797-6_5

1 Introduction

It is unrevealed that fossil fuels are the most prominent source of energy worldwide and gain the attention of millions of users in various applications in different sectors, for example, business, transportation, research, and industries (Abengoa Bioenergy (2011). Presently, three main forms of fossil fuels are available in the market; coal, oil, and natural gas (Achten et al. 2008). However, there are many challenges faced by societies, for example, energy security, increase of oil price, resource depletion, and climate change that escort researchers towards the search for new, attractive and renewable sources of energy which can replace the fossil fuels (Acikgoz and Kockar 2009). Among presently available renewable energy options, biofuel is one of the potential scopes (Atabani et al. 2012). Currently, three main types of biofuels have been discussed broadly by many researches including biodiesel, bioethanol, and biohydrogen (Atadashi et al. 2010). There are many efforts and research focused to improve the quality of a biofuel which is feasible from a commercial point of view (Ates and Isakdag 2008). Economical production of biofuels can be done by using various renewable biological sources, such as lignocellulosic biomass, algae, soybean, jatropha, corn, palm, coconut, rice bran, linseed, jojoba, castor, etc. The 2010 global annual production of agricultural residues was around 5.1 billion dry ton (Atsumi et al. 2008).

The waste generated by the agricultural, forestry, and aquaculture sectors is increasing with the increasing population and thus the waste from this sector will be increasing further in the future (Bals et al. 2007). Besides cost, development of sustainable energy technologies from renewable feedstocks has become more important to diminish global climate change (Bastian et al. 2011). To date, more research and process development have focused on the production of biofuels from sugar content obtained from bioconversion of lignocellulosic feedstocks (Belyea et al. 2004). Apart from sugars, proteins are also main components in waste biomass. For instance, the protein content reached up to ~16% of the dry weight in lignocellulosic biomass and was considered as by-product after the pretreatment of these biomasses (Boateng et al. 2009). Biofuels process using cellulosic biomass, resultant accumulation of these protein by-products, but there are no strategies for conversion of these by-products into fuels. Further, these reduced nitrogen present in form of protein is not required to recycle to be used in biofuels production as nitrogen source and production of biofuels from nitrogen source will not contribute to greenhouse gas (Boateng et al. 2008). Moreover, recycling of ammonia from the protein containing biomass as a fertilizer for photosynthetic feedstocks may be able to close the nitrogen cycle. Henceforth, protein utilization in a controlled manner can allow for the recycling of ammonia.

2 Biofuels Production Processes and Sustainable Use of Plant Biomass

Biofuels production for transport is a potential alternative to replace conventional fuels such as gasoline and diesel and has gained more attention in the past few years (Boateng et al. 2008). Additionally, both conventional biofuels (biofuels produced from edible crops) and advanced biofuels (biofuels produced from nonedible crops) can be easily differentiated. Advanced biofuels include cellulosic ethanol, biohydrogen, butanol as well as thermochemically and catalytically produced Fischer-Tropsch diesel and synthetic natural gas (SNG) (Bok et al. 2012). However, the cost-intensive production costs of these advanced biofuels make them limited for commercial applications (Borugadda and Goud 2012). To overcome these bottlenecks, continues efforts (such as the use of cheap, carbon-rich and easily available raw materials) are going on. Plant residues are potential substrates for these biofuels production process and contain high percentage of cellulose. These cellulosic biomasses are easily converted into biofuels (biohydrogen or bioethanol) via enzymatic hydrolysis and fermentation process. The production of value-added bio-based products using these residues such as biofuels and chemicals is important to maximize full biomass-to-products value chains and effectively lowers the production cost of the biofuels market (Nayan et al. 2013).

3 Utilization of Protein-Rich Residues for Biofuel Production

Plant biomass, enzymes used in the process of biomass conversion, and fermentation via microorganisms are the way from where nitrogen-rich biofuel production residuals come (Onay and Koçkar 1921). The common name of this material by which they are sold in market is known as distiller's dried grains with soluble (DDGS) (Onay 2007). The percentage amount of DDGS is varied among different plant residues, for example, first-generation biofuels production, such as maize, wheat, sweet sorghum sugar beet, and sugarcane, typically contain between 20 and 40% protein (Ong et al. 2011; Özbay et al. 2008). Additionally, the leaves of crops, for example, alfalfa and cassava also contain ~20–40% protein. In contrast, the percentage of protein is very low in hydrolysate of lignocellulosic biomass such as wheat straw and maize stover (Parnaudeau et al. 2008; Phukan et al. 2011). Nevertheless, some recent studies also suggested a specific model to recover these proteins via two-step method to extract the proteins with warm aqueous ammonia (Phukan et al. 2011). In this two-step method, the protein solution is mechanically dewatered, for recovery of ammonia followed by drying the protein solution. Moreover, around 84% of protein can be recovered using this method. In this series, proteins by-products are obtained from oil or biodiesel production, for example, the seedcakes of rapeseed, soybean, and Jatropha. The variations in the amount of protein in Jatropha seedcake are large and may be dependent on the method of production of the seedcake. In addition, amount of protein in these plants are varied around 25–58% (Pütün et al. 2004, 2005, 2006).

4 Current Biomass Availability

4.1 Sugarcane and Sugar Beet By-products

Sugarcane is used as one of the potential crops for bioethanol production and contains an average crude protein mass fraction of 13% of the dry matter (Table 1). A typical sugarcane plant contains 72% water and 11 kt of the residual protein in its mass fraction (Ro et al. 2006; San et al. 2002; Şen and Kar 2011). In one of the studies by Deepchand et al., isolation of the protein from leaves and tops of sugarcane was reported. Authors calculated in their experiment that 100 kWh electricity and 8.3 dm^3 ethanol along with 3 kg proteins can be produced from every ton of fresh leaves and tops of sugarcane (Sensoz and Angin 2008; Shahid and Jamal 2011). The main product of their study was electricity whereas ethanol and protein were the by-products. Based on this study, sugarcane can be a potential source for the production of protein by-products.

The annual sugar beet leaf production is approximately 140 Mt worldwide. The major part of the fresh leaves contains water, a mass fraction ~86.4%, and around 3.2% protein. That means an annual total leaf protein potential of 4.5 Mt worldwide. Moreover, the production plant of bioethanol of British Sugar in Wissington, UK, showed a capacity for 70 dam^3 of bioethanol per year, thus using 110 kt of sugar coming from 650 kt sugar beets (Shen and Liao 2008; Shen et al. 2009; Singh and Singh 2010).

4.2 Wheat and Cassava By-products

Bioethanol production from wheat is also carried out at large scale and it was estimated that ~0.5 hm^3 of ethanol could be produced from ~0.36 Mt of wheat DDGS (Distillers Dried Grains with Solubles) which is equivalent to 0.12 Mt of crude protein (Spiehs et al. 2002). Furthermore, the annual potential of 224 Mt of fresh cassava leaves has ~15.5 Mt crude protein (Lammens et al. 2012). Cassava is used for bioethanol production and it was estimated that 84–89 kg of cassava sludge produce around 333 dm^3 ethanol (Tewe 2004). Moreover, at the end of the reaction, a cassava root contains 1–2% protein in a mass fraction (Titiloye et al. 2013) and the amount of protein present in the sludge is a mass fraction of ~35%, thus from 10 dm^3 cassava depended ethanol, 1 kg protein is formed in a protein-rich sludge (Tröger et al. 2013).

Table 1 Compositional analysis of different plant biomass including lignocellulosic biomass[a]

S. No.	Name of DDGS of crops	Mass fraction of crude protein (%)	References
1.	Maize	~30	Niu and Sun (2012)
2.	Wheat	~35	Vassilev et al. (2012)
3.	Sweet Sorghum	~35	Wang et al. (2009)
4.	Sugarbeet	~32	Wang et al. (2011)
5.	Sugarcane	~25	Wei et al. (2006)
6.	Cassava	~10	Werther et al. (2000)
7.	Alfaalfa	~40	Widyaratne and Zijlstra (2007)
8.	Jatropha Seed	~50	Yang et al. (2004)
9.	Soybean meal	~60	Yorgun et al. (2001)
10.	Sunflower seed meal	~40	Zheng (2008)

[a]Atsumi et al. (2008), Bals et al. (2007), Bastian et al. (2011), Belyea et al. (2004), Boateng et al. (2009, 2008), Bok et al. (2012), Borugadda and Goud (2012), Brebu et al. (2010), Bridgeman et al. (2007), British Sugar Plc (2010), Celma et al. (2007), Corredor et al. (2006), Dale et al. (2009), Dale and Matsuoka (1981), Das and Ganesh (2003), Deepchand (1985), Deike and Mol (1996), Demiral and Ayan (2011), Demiral and Şensöz (2008), Demirbas (2007, 2009, 2010), DeSisto et al. (2010), Dong et al. (1987), Duke (2011), Duman et al. (2011), Ertaş and Hakkı Alma (2010), FAOSTAT (2010), Freire et al. (2001), Garcia-Perez et al. (2008), Gerpen (2005), Gu et al. (2013), Halim (2012), Hartemink (2008), Hong et al. (2014), Kim et al. (2013), Lamsen and Atsumi (2012), Lan and Liao (2012), Lazazzera (2000), Lestari et al. (2010), Lewicki (2001), Li et al. (2004, 2012), López et al. (2013), Luo et al. (2010), Makkar et al. (2008), Merodio and Sabater (1988), Min et al. (2013), Mourant et al. (2013), Murata et al. (2011), Naik et al. (2010)

5 Limitations and Future Suggestions

The main focus of present chapter was to review the current protein availability from cellulose-rich lignocellulosic plant biomass, in order to obtain biofuels production and significant utilization of its protein by-product (Tsai et al. 2006). Protein obtained from pretreatment process of lignocellulosic biomass may be used as media component for solid state/submerge fermentation. Thus, the overall yields of the original feedstocks may enhance when these protein residues consumed by fermentation mediated microorganisms (Tsai et al. 2012). Moreover, this process also helps to reduce waste and environmental pollution as protein fermentation eliminates DDGS by-products which are used as animal feed and thus prevents imbalance the nitrogen cycle (Tushar et al. 2010).

Although these carbohydrate and protein-rich lignocellulosic residues are the major income source for biofuel production, they are suffering a number of drawbacks. Lack of development of technology in sufficient utilization of protein by-product obtained from this lignocellulosic biomass is considered as one of the major hurdles in developing area (Urriola et al. 2009; Uzun et al. 2007; Raveendran et al. 1995; Silitonga et al. 2011). Though, present market of animal feed is able to absorb

DDGS produced by the biofuel industry, it is still not certain whether the market will be continued to produce biofuel using protein rich source either alone or with the combination of carbohydrate rich sources. Additionally, residues obtained from lignocellulosic biofuel production process contain much more complex mixtures thus it is uncertain, whether such types of residues can be effectively digested by animals (Uzun et al. 2006). A feasible scheme is to deamination of the protein residues into ammonia, which can be further utilized as fertilizer for plant growth as well as nitrogen source for fermentation medium. Use of other plant residues which contain high amount of protein such as seed cakes and pulses residues may play potential role in the utilization of these kinds of protein by-products.

6 Conclusion

The aim of this chapter was to assess the availability and utilization of protein by-product in lignocellulosic plant residues available for biofuels production. The study showed that there is enough protein present in lignocellulosic biomass which can be used for cost-effective biofuels production and can replace the fossil fuel market. These protein rich plant residues can be obtained as the byproduct after the pre-treatment (of biomass) and can be efficiently converted in to biofuels along with pretreated lignocellulosic biomass. Although it is growing area of research and very few reports are available specifically on this, more studies are needed towards also quantification of the protein availability and technology for its complete utilization. The huge availability of lignocellulosic biomass and developing technology for utilization of carbohydrates and protein into biofuels may play a significant role in lowering the production cost of biofuels.

Acknowledgements Authors N. S. thankfully acknowledge IIT (BHU) Varanasi, India for providing Institute Post-Doctoral fellowship. P. K. Mishra and N. S., acknowledge Department of Chemical Engineering and Technology IIT (BHU) Varanasi. M.S. acknowledges the Department of Science and Technology, Govt. of India for DST INSPIRE Faculty [IFA13-MS-02] 2014 award.

References

Abengoa Bioenergy (2011) Abengoa europoort bioethanol plant information. Abengoa Bioenergy, S.A, Madrid, c2011 [cited 2012 April 7]. Available from: http://www.abengoabioenergia.com/export/sites/abg_bioenergy/resources/pdf/acerca_de/en/Annual_report_2011_3.pdf

Achten WMJ, Verchot L, Franken YJ, Mathijs E, Singh VP, Aerts R et al (2008) Jatropha bio-diesel production and use. Biomass Bioenergy 32(12):1063–1084

Acikgoz C, Kockar OM (2009) Characterization of slow pyrolysis oil obtained from linseed (Linum usitatissimum L.). J Anal Appl Pyrol 85(1–2):151–154

Atabani AE, Silitonga AS, Badruddin IA, Mahlia TMI, Masjuki HH, Mekhilef SA (2012) Comprehensive review on biodiesel as an alternative energy resource and its characteristics. Renew Sustain Energy Rev 16(4):2070–2093

Atadashi IM, Aroua MK, Aziz AA (2010) High quality biodiesel and its diesel engine application: are view. Renew Sustain Energy Rev 14(7):1999–2008
Ates F, Isakdag MA (2008) Evaluation of the role of the pyrolysis temperature in straw biomass samples and characterization of the oils by GC/MS. Energy Fuels 22(3):1936–1943
Atsumi S, Cann AF, Connor MR, Shen CR, Smith KM, Brynildsen MP, Chou KJ, Hanai T, Liao JC (2008) Metabolic engineering of Escherichia coli for 1-butanol production. Metab Eng 10:305–311. https://doi.org/10.1016/j.ymben.2007.08.003
Bals B, Teachworth L, Dale B, Balan V (2007) Extraction of proteins from switchgrass using aqueous ammonia within an integrated biorefinery. Appl Biochem Biotech 143(2):187–198
Bastian S, Liu X, Meyerowitz JT, Snow CD, Chen MM, Arnold FH (2011) Engineered ketol-acid reductoisomerase and alcohol dehydrogenase enable anaerobic 2-methylpropan-1-ol production at theoretical yield in Escherichia coli. Metab Eng 13:345–352
Belyea RL, Rausch KD, Tumbleson ME (2004) Composition of corn and distillers dried grains with solubles from dry grind ethanol processing. Bioresour Technol 94(3):293–298
Boateng AA et al (2008) Production of bio-oil from alfalfa stems by fluidized-bed fast pyrolysis. Ind Eng Chem Res 47(12):4115–4122
Boateng AA et al (2009) Energy-dense liquid fuel intermediates by pyrolysis of guayule (*Parthenium argentatum*) shrub and bagasse. Fuel 88(11):2207–2215
Bok JP et al (2012) Fast pyrolysis of coffee grounds: characteristics of product yields and biocrude oil quality. Energy 47(1):17–24
Borugadda VB, Goud VV (2012) Biodiesel production from renewable feedstocks: status and opportunities. Renew Sustain Energy Rev 16(7):4763–4784
Brebu M et al (2010) Co-pyrolysis of pine cone with synthetic polymers. Fuel 89(8):1911–1918
Bridgeman TG et al (2007) Influence of particle size on the analytical and chemical properties of two energy crops. Fuel 86(1–2):60–72
British Sugar Plc (2010) British sugar, bioethanol. British Sugar Plc; c2010, London [cited 2012 April 7]. Available from: http://www.britishsugar.co.uk/Bioethanol.aspx
Celma AR, Rojas S, Lopez-Rodriguez F (2007) Waste-to-energy possibilities for industrial olive and grape by-products in extremadura. Biomass Bioenergy 31(7):522–534
Corredor DY, Bean SR, Schober T, Wang D (2006) Effect of decorticating sorghum on ethanol production and composition of DDGS. Cereal Chem 83(1):17–21
Dale BE, Matsuoka M (1981) Protein recovery from leafy crop residues during biomass refining. Biotechnol Bioeng 23(6):1417–1420
Dale BE, Allen MS, Laser M, Lynd LR (2009) Protein feeds coproduction in biomass conversion to fuels and chemicals. Biofuel Bioprod Biorefin 3(2):219–230
Das P, Ganesh A (2003) Bio-oil from pyrolysis of cashew nut shell—a near fuel. Biomass Bioenergy 25(1):113–117
Deepchand K (1985) System for the production of electricity, leaf protein and single cell protein from sugar cane tops and leaves. Sol Energy 35(6):477–482
Deike V, Mol JD (1996) Raw material compendium, 2 edn. Brussels Novus International Inc
Demiral İ, Ayan EA (2011) Pyrolysis of grape bagasse: effect of pyrolysis conditions on the product yields and characterization of the liquid product. Biores Technol 102(4):3946–3951
Demiral İ, Şensöz S (2008) The effects of different catalysts on the pyrolysis of industrial wastes (olive and hazelnut bagasse). Biores Technol 99(17):8002–8007
Demirbas A (2007) The influence of temperature on the yields of compounds existing in bio-oils obtained from biomass samples via pyrolysis. Fuel Process Technol 88(6):591–597
Demirbas A (2009) Progress and recent trends in biodiesel fuels. Energy Convers Manag 50(1):14–34
Demirbas A (2010) Use of algae as biofuel sources. Energy Convers Manage 51(12):2738–2749
DeSisto WJ et al (2010) Fast pyrolysis of pine sawdust in a fluidized-bed reactor. Energy Fuels 24(4):2642–2651
Dong FM, Rasco BA, Gazzaz SS (1987) A protein quality assessment of wheat and corn distillers' dried grains with solubles. Cereal Chem 64(5):327–332

Duke JA (2011) Handbook of energy crops [Internet]. Center for New Crops & Plant Products, Purdue University; c2011, West Lafayette [cited 2012 April 7]. New CROP, Crop Index. Available from: http://www.hort.purdue.edu/newcrop/Indices/index_ab.html

Duman G et al (2011) The slow and fast pyrolysis of cherry seed. Biores Technol 102(2):1869–1878

Ertaş M, Hakkı Alma M (2010) Pyrolysis of laurel (*Laurus nobilis* L.) extraction residues in a fixed-bed reactor: characterization of bio-oil and bio-char. J Anal Appl Pyrol 88(1):22–29

FAOSTAT (2010) Production, crops, production quantity. Food and Agriculture Organization of the United Nations; 2010, Rome [cited 2012 April 17]. Available from: http://faostat.fao.org, [Files updated annually]

Freire WJ, Rolim MM, Cortez LAB, Bauen A (2001) Use of sugarcane ethanol vinasse for brick manufacture. Agric Mech Asia Afr Lat Am 32(1):51–54

Garcia-Perez M et al (2008) Fast pyrolysis of oil mallee woody biomass: effect of temperature on the yield and quality of pyrolysis products. Ind Eng Chem Res 47(6):1846–1854

Gerpen JV (2005) Biodiesel processing and production. Fuel Process Technol 86(10):1097–1107

Gu S et al (2013) A detailed study of the effects of pyrolysis temperature and feedstock particle size on the preparation of nanosilica from rice husk. Ind Crops Prod 50:540–549

Halim R, Danquah MK, Webley PA (2012) Extraction of oil from microalgae for biodiesel production: a review. Biotechnol Adv 30(3):709–732

Hartemink AE (2008) Sugarcane for bioethanol: soil and environmental issues (Chapter 3). Adv Agron 99:125–182

Hong T, Koo C, Kwak T, Park HS (2014) An economic and environmental assessment for selecting the optimum new renewable energy system for educational facility. Renew Sustain Energy Rev 29:286–300

Kim SW et al (2013) Bio-oil from the pyrolysis of palm and Jatropha wastes in a fluidized bed. Fuel Process Technol 108:118–124

Lammens TM, Franssen MCR, Scott EL, Sanders JPM (2012) Availability of protein-derived amino acids as feedstock for the production of bio-based chemicals. Biomass Bioenergy 44:168–181

Lamsen EN, Atsumi S (2012) Recent progress in synthetic biology for microbial production of C3-C10 alcohols. Front Microbiol 3:196–201

Lan EI, Liao JC (2012) ATP drives direct photosynthetic production of 1-butanol in cyanobacteria. Proc Natl Acad Sci USA 109:6018–6023

Lazazzera BA (2000) Quorum sensing and starvation: signals for entry into stationary phase. Curr Opin Microbiol 3:177–182

Lestari D, Mulder W, Sanders J (2010) Improving Jatropha curcas seed protein recovery by using counter current multistage extraction. Biochem Eng J 50(1–2):16–23

Lewicki W (2001) An introduction to vinasse from beet and cane molasses fermentation. Int Sugar J 103(1227):126–128

Li S et al (2004) Fast pyrolysis of biomass in free-fall reactor for hydrogen-rich gas. Fuel Process Technol 85(8–10):1201–1211

Li D et al (2012) Preparation and characteristics of bio-oil from the marine brown alga *Sargassum patens* C. Agardh. Bioresour Technol 104:737–742

López FA et al (2013) Textural and fuel characteristics of the chars produced by the pyrolysis of waste wood, and the properties of activated carbons prepared from them. J Anal Appl Pyrol 104:551–558

Luo S et al (2010) Effect of particle size on pyrolysis of single-component municipal solid waste in fixed bed reactor. Int J Hydrogen Energy 35(1):93–97

Makkar HPS, Francis G, Becker K (2008) Protein concentrate from Jatropha curcas screw-pressed seed cake and toxic and antinutritional factors in protein concentrate. J Sci Food Agr 88(9):1542–1548

Merodio C, Sabater B (1988) Preparation and properties of a white protein fraction in high yield from sugar beet (Beta vulgaris) leaves. J Sci Food Agr 44(3):237–243

Min Z et al (2013) Catalytic reforming of tar during gasification. Part III. Effects of feedstock on tar reforming using ilmenite as a catalyst. Fuel 103:950–955

Mourant D et al (2013) Effects of temperature on the yields and properties of bio-oil from the fast pyrolysis of mallee bark. Fuel 108:400–408

Murata K et al (2011) Analyses of liquid products from catalytic pyrolysis of jatropha seed cakes. Energy Fuels 25(11):5429–5437

Naik S et al (2010) Characterization of Canadian biomass for alternative renewable biofuel. Renew Energy 35(8):1624–1631

Nayan NK, Kumar S, Singh RK (2013) Production of the liquid fuel by thermal pyrolysis of neem seed. Fuel 103:437–443

Onay O (2007) Influence of pyrolysis temperature and heating rate on the production of bio-oil and char from safflower seed by pyrolysis, using a well-swept fixed-bed reactor. Fuel Process Technol 88(5):523–531

Onay O, Koçkar OM (2006) Pyrolysis of rapeseed in a free fall reactor for production of bio-oil. Fuel 85(12–13):1921–1928

Ong HC, Mahlia TMI, Masjuki HH, Norhasyima RS (2011) Comparison of palm oil, Jatropha curcas and Calophylluminophyllum for biodiesel: a review. Renew Sustain Energy Rev 15(8):3501–3515

Özbay N et al (2008) Characterization of bio-oil obtained from fruit pulp pyrolysis. Energy 33(8):1233–1240

Parnaudeau V, Condom N, Oliver R, Cazevieille P, Recous S (2008) Vinasse organic matter quality and mineralization potential, as influenced by raw material, fermentation and concentration processes. Bioresour Technol 99(6):1553–1562

Phukan MM et al (2011) Microalgae Chlorella as a potential bio-energy feedstock. Appl Energy 88(10):3307–3312

Pütün AE, Apaydin E, Pütün E (2004) Rice straw as a bio-oil source via pyrolysis and steam pyrolysis. Energy 29(12–15):2171–2180

Pütün AE et al (2005) Bio-oil from olive oil industry wastes: pyrolysis of olive residue under different conditions. Fuel Process Technol 87(1):25–32

Pütün E, Uzun BB, Pütün AE (2006) Fixed-bed catalytic pyrolysis of cotton-seed cake: effects of pyrolysis temperature, natural zeolite content and sweeping gas flow rate. Bioresour Technol 97(5):701–710

Raveendran K, Ganesh A, Khilar KC (1995) Influence of mineral matter on biomass pyrolysis characteristics. Fuel 74(12):1812–1822

Ro DK, Paradise EM, Ouellet M, Fisher KJ, Newman KL, Ndungu JM, Ho KA, Eachus RA, Ham TS, Kirby J, Chang MCY, Withers ST, Shiba Y, Sarpong R, Keasling JD (2006) Production of the antimalarial drug precursor artemisinic acid in engineered yeast. Nature 440:940–943

San KY, Bennett GN, Berrios-Rivera SJ, Vadali RV, Yang YT, Horton E, Rudolph FB, Sariyar B, Blackwood K (2002) Metabolic engineering through cofactor manipulation and its effects on metabolic flux redistribution in Escherichia coli. Metab Eng 4:182–192

Şen N, Kar Y (2011) Pyrolysis of black cumin seed cake in a fixed-bed reactor. Biomass Bioenerg 35(10):4297–4304

Sensoz S, Angin D (2008) Pyrolysis of safflower (*Charthamus tinctorius* L.) seed press cake in a fixed-bed reactor: part 2. Structural characterization of pyrolysis bio-oils. Biores Technol 99(13):5498–5504

Shahid EM, Jamal Y (2011) Production of biodiesel: a technical review. Renew Sustain Energy Rev 15(9):4732–4745

Shen CR, Liao JC (2008) Metabolic engineering of Escherichia coli for 1-butanol and 1-propanol production via the keto-acid pathways. Metab Eng 10:312–320

Shen DK et al (2009) Kinetic study on thermal decomposition of woods in oxidative environment. Fuel 88(6):1024–1030

Silitonga AS, Atabani AE, Mahlia TMI, Masjuki HH, Badruddin IA, Mekhilef S (2011) A review on prospect of Jatropha curcas for biodiesel in Indonesia. Renew Sustain Energy Rev 15(8):3733–3756

Singh SP, Singh D (2010) Biodiesel production through the use of different sources and characterization of oils and their esters as the substitute of diesel: a review. Renew Sustain Energy Rev 14(1):200–216
Spiehs MJ, Whitney MH, Shurson GC (2002) Nutrient database for distiller's dried grains with solubles produced from new ethanol plants in Minnesota and South Dakota. J Anim Sci 80(10):2639–2645
Tewe OO (2004) Cassava for livestock feed in sub-Saharan Africa. The global cassava development strategy. Food and Agricultural Organization of the United Nations
Titiloye JO, Abu Bakar MS, Odetoye TE (2013) Thermochemical characterisation of agricultural wastes from West Africa. Ind Crops Prod 47:199–203
Tröger N, Richter D, Stahl R (2013) Effect of feedstock composition on product yields and energy recovery rates of fast pyrolysis products from different straw types. J Anal Appl Pyrol 100:158–165
Tsai WT, Lee MK, Chang YM (2006) Fast pyrolysis of rice straw, sugarcane bagasse and coconut shell in an induction-heating reactor. J Anal Appl Pyrol 76(1–2):230–237
Tsai W-T, Liu S-C, Hsieh C-H (2012) Preparation and fuel properties of biochars from the pyrolysis of exhausted coffee residue. J Anal Appl Pyrol 93:63–67
Tushar MSHK et al (2010) Analysis of gaseous and liquid products from pressurized pyrolysis of flax straw in a fixed bed reactor. Ind Eng Chem Res 49(10):4627–4632
Uçar S, Karagöz S (2009) The slow pyrolysis of pomegranate seeds: the effect of temperature on the product yields and bio-oil properties. J Anal Appl Pyrol 84(2):151–156
Urriola PE, Hoehler D, Pedersen C, Stein HH, Shurson GC (2009) Amino acid digestibility of distillers dried grains with solubles, produced from sorghum, a sorghum-corn blend, and corn fed to growing pigs. J Anim Sci 87(8):2574–2580
Uzun BB, Pütün AE, Pütün E (2006) Fast pyrolysis of soybean cake: product yields and compositions. Bioresour Technol 97(4):569–576
Uzun BB, Pütün AE, Pütün E (2007) Composition of products obtained via fast pyrolysis of olive-oil residue: effect of pyrolysis temperature. J. Anal Appl Pyrol 79(1–2):147–153
Van Dien SJB AP, Haselbeck R, Pujol-baxely CJ, Niu, W, Trawick JD, Burk MJ, Osterhout RE, Sun J (2012) Microorganisms for the production of 1,4-butanediol and related methods. Genomatica, Inc US Patent Application 20120225463
Vassilev SV et al (2012) An overview of the organic and inorganic phase composition of biomass. Fuel 94:1–33
Wang Y, Tilley M, Bean S, Sun XS, Wang D (2009) Comparison of methods for extracting kafirin proteins from sorghum distillers dried grains with solubles. J Agr Food Chem 57(18):8366–8372
Wang L, Shahbazi A, Hanna MA (2011) Characterization of corn stover, distiller grains and cattle manure for thermochemical conversion. Biomass Bioenergy 35(1):171–178
Wei L et al (2006) Characteristics of fast pyrolysis of biomass in a free fall reactor. Fuel Process Technol 87(10):863–871
Werther J et al (2000) Combustion of agricultural residues. Prog Energy Combust Sci 26(1):1–27
Widyaratne GP, Zijlstra RT (2007) Nutritional value of wheat and corn distiller's dried grain with solubles: digestibility and digestible contents of energy, amino acids and phosphorus, nutrient excretion and growth performance of growerfinisher pigs. Can J Anim Sci 87(1):103–114
Yang HP et al (2004) Thermogravimetric analysis-fourier transform infrared analysis of palm oil waste pyrolysis. Energy Fuels 18(6):1814–1821
Yorgun S, Sensoz S, Kockar OM (2001) Characterization of the pyrolysis oil produced in the slow pyrolysis of sunflower-extracted bagasse. Biomass Bioenergy 20(2):141–148
Zheng JL (2008) Pyrolysis oil from fast pyrolysis of maize stalk. J Anal Appl Pyrol 83(2):205–212

Biofuels from Microorganisms

Mariam Amer, AbdelGawad Saad and Nahed K. Ismail

Abstract Biofuel as a renewable energy, can be produced from many resources, but the easiest, safest, and most economic resources used are organisms—natural materials like algae—especially microscopic organisms. Microalgae are characterized by their ability to be grown both naturally and quickly, and represent a source of carotenoids, lipids, and polysaccharides. *Chlamydomonas reinhardtii*, *Dunaliella salina*, and various *Chlorella* species permit the extraction of about 5–7% biodiesel from their cells. Producing bioethanol to a higher concentration of 60% can be obtained using *Chlorococum* sp. The best technique for using microalgae to produce biofuel as biodiesel and bioethanol is a biochemical technique, that is, the photo-fermentation technique used to produce biohydrogen. The biochemical technique uses a process known as pyrolysis in which biomass is heated, in the absence of air, to temperatures above 500 °C for short periods (a few minutes). Also, *C. reinhardtii* can generate high condensation levels of biohydrogen. To produce biohydrogen, a quick fermentation process is required using non-sulfur bacteria, with light as an energy source , to produce organic acids by dark fermentation.

M. Amer · A. Saad (✉) · N. K. Ismail
Bio-system Engineering Department, Agricultural Engeenering Research Institute (AEnRI), Agricultural Research Center (ARC), Giza, Egypt
e-mail: en_gawad2000@yahoo.com

N. Srivastava et al. (eds.), *Sustainable Approaches for Biofuels Production Technologies*, Biofuel and Biorefinery Technologies 7,
https://doi.org/10.1007/978-3-319-94797-6_6

Graphical Abstract

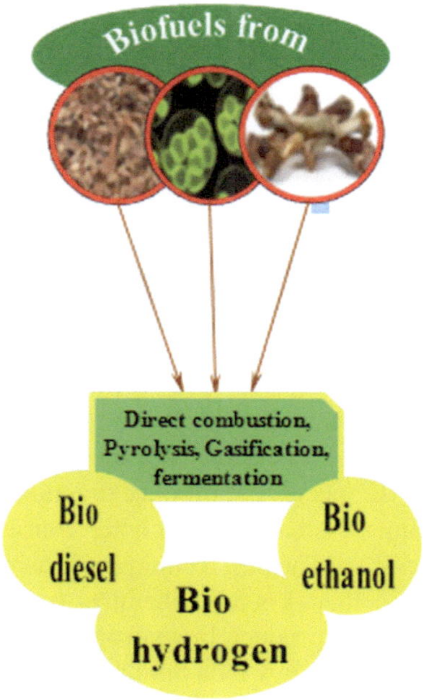

Keywords Biofuel · Biodiesel · Biohydrogen · Bioethanol · Macroalgae Microalgae

1 Introduction

Due to the continuous decline in natural energy sources and the fact that most sources are harmful to the environment, an attempt is being made to search for other sources of energy which are safer, faster, and more productive. The most important of these sources, coming from natural materials, is called biomass. Biomass can be used as the raw material for bioenergy derived from biodiesel, bioethanol, biogas, and biobutane. These are collectively known as biofuels. These biofuels can be produced mainly from the materials within hemicellulose, cellulose, and lignins. Biomass sources include bones, wood, maize, grass, algae, oils, etc.

Fossil fuels damage the environment by producing greenhouse gas emissions leading to an enhanced global warming (Chisti 2007; Medipally et al. 2015). The negative effects of using fossil fuels are: reducing fossil fuel reserves; diminishing available resources leading to increasing CO_2 concentrations in the atmosphere causing a change in climate; and geopolitical strife. Therefore, the greatest challenge is searching for "clean" energy resources (Mata et al. 2010; Medipally et al.

2015; Shuba and Kifle 2018). In addition, the provision of multiple sources of energy protects against human conflicts driven by, for example, the demand for oil.

Biofuels are classified into four generations based on their production technologies: first-generation fuels which are made from vegetable oils, starch, sugar, or animal fats; second-generation fuels which are made from corn, wheat straw, non-food crops, wood, or solid waste; third-generation fuels which are made from algae; and finally fourth-generation fuels made from the conversion of vegetable oils and biodiesel for biogasoline. Giving particular attention to third-generation biofuels, algal biomass can accumulate considerably high amounts of lipids comparing with the biomass of oil plants (Abdelaziz et al. 2013; Voloshin et al. 2016).

Algae grows naturally and quickly and produces oxygen by photosynthesis. In addition, macroalgae does not require land, so there is no competition between algae and plants in terms of space. Therefore, macroalgae biofuels have little effect on farms or food supplies and do not require compound treatment methods as compared with lignocellulose-enriched biomass (Voloshin et al. 2016).

The main difference between algae bioenergetics and plant bioenergetics is the technology used to increase biomass. Plant bioenergetics requires the utilization of valuable resources and provides a relatively low yield in terms of the proportion of the organic feedstock mass to the mass of the biofuel synthesized. On the other hand, plants do not require any additional production methods, besides the standard growth techniques already used in agriculture and the creation of specific growing conditions. Microalgae can grow in conditions which are unsuitable for plant growth, that is, saline soils, wastewater, etc. (Chisti 2007; Wang et al. 2008).

Phytoplankton or microalgae are commonly found in oceans: the most well known being dinoflagellates, diatoms, green algae, and blue–green algae. The most important resource for carotenoids, lipids, and polysaccharides are marine unicellular microalgae, which have been extensively studied in the scope of biofuel production and fodder supplements (Liau et al. 2010). Furthermore, land plants can realize a photoconversion productivity of less than 1% in temperate climates, whereas microalgae can convert 5% of solar energy into chemical energy (Rösch et al. 2012).

2 Energy Conversion Process

The actual conversion of biomass into biofuels comes after its cultivation and preparation processes (Chisti 2007; Nigam and Singh 2011).

There are different techniques for converting biomass to energy. The first uses a chemical technique (hydrolysis and/or transesterification) which provides certain reactions in the presence of a catalyst. The second is a biochemical technique (fermentation and/or hydrolysis) which depends on the nature of the chemical processes which occur in living cells. Direct combustion represents a third technique where heat energy is converted to electrical energy. The last technique is called thermo-chemical (gasification, pyrolysis, and liquefaction) and includes treatments of feedstock, under

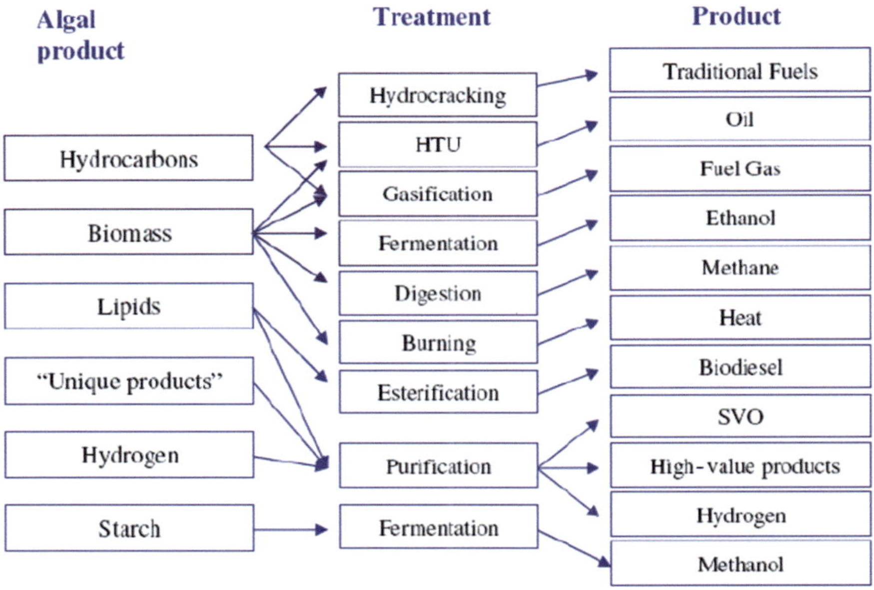

Fig. 1 Different of methods used to produce biofuel from algae (Voloshin et al. 2016). HTU—Hydro thermal upgrading, SVO—Straight vegetable oil

high pressures and temperatures, to obtain compounds at both low O_2 content and molecular weight.

There are different methods used to produce algal biofuels—Fig. 1 illustrates these methods (Voloshin et al. 2016).

3 Biodiesel

One of the methods of multiplying the production capacity is the production of biodiesel. Biodiesel is one alternative fuel which is obtained by a transesterification reaction in the presence of triglyceride oil and monohydric alcohols. Biodiesel is non-toxic, technologically sensible, and biodegradable when it is obtained from renewable resources. It can be obtained from residues of vegetable oil, fish oil, chicken fat, and algal oil (Lang et al. 2002; Spolaore et al. 2006; Sharif et al. 2007) which therefore partly decreases our dependency on oil-based fossil fuels.

Algae (macro and micro) generally has a greater photosynthetic effect than other biomass. Algae is the many sources of bio-diesel and the highest supply from feedstock for biodiesel. It can produce more than 250 times the oil produced per acre of soybeans. Similarly, algae produces 7 to 31 times more biodiesel than palm oil. It is thought that biodiesel automotive fuel, produced from algae, could be used to replace gasoline. The preference is to use microalgae rather than macroalgae to pro-

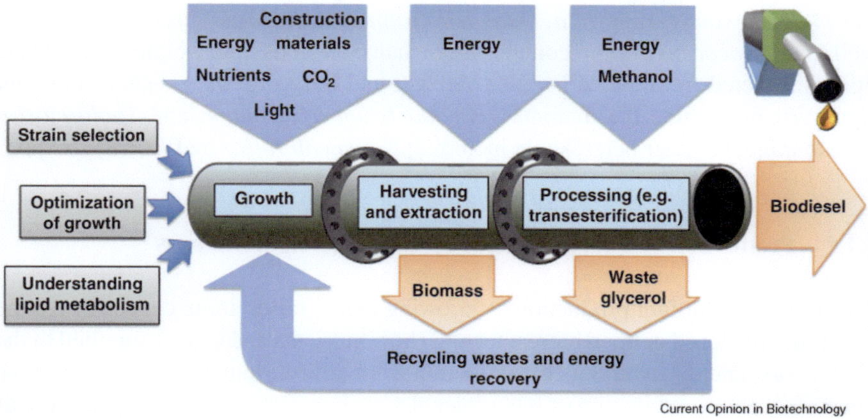

Fig. 2 Processing algal biofuel in a pipeline (Scott et al. 2010)

duce biodiesel. Microalgae are categorized as organisms which are less than 2 mm in diameter and are capable of photosynthesis. Microalgae is characterized by the fact that it is faster and easier to grow and produces greater oil yields than macroalgae. Macroalgae is not usually used to produce biodiesel. However, biodiesel can be produced from macroalgae if it contains a lower lipid content than microalgae. Some researchers concluded that using *Oedogonium* sp. allowed them to obtain higher contents of biodiesel than *Spirogyra* sp. (Hossain and Salleh 2008).

Some polyunsaturated organisms may contain a high amount of fatty acid as docosahexaenoic acid (DHA), which consists of 22 atoms of carbon in 6 dualties which belong to the so-called w-3 group. These organisms can be grown without light on heterotrophically organic substrates. These types of macroalgae contain from 1.3 to 7.8% dw of lipid (Sijtsma and Swaaf 2004).

3.1 *Producing Biodiesel from Algae*

Figure 2 illustrates the major stages which must be taken into consideration. There are many factors which must be optimized such as material inputs (nutrients and growth energy for homogenous), energy, suitable treatment of spent media, residue products, and residual biomass (Scott et al. 2010).

3.1.1 Algal Strain Selection

One vital consideration is algal strain selection. Algae is a non-flowering plant as an aquatic organism which feeds by photosynthesis. Approximately 300,000 species of algae have been identified from different sources. Some species of green algae such as *Chlamydomonas reinhardtii*, *Dunaliella salina*, and various *Chlorella* species, as well as *Botryococcus braunii* contains more than 60% of its lipid weight. Much of the lipid is secreted from the cell walls (Metzger and Largeau 2005). Other important algae groups include the diatoms *Phaeodactylum tricornutum* and *Thalassiosira pseudonana* and other heterokonts including *Nannochloropsis* and *Isochrysis* spp.

3.1.2 Production of Fuel Molecules by Growing Algal Biomass

The process of biofuel production by growing algae, raises some concerns: (1) the feasibility of closed or open bioreactors, (2) how nutrients and CO_2 is supplied to the bioreactors, and (3) avoiding contamination from adventitious organisms. For most microalgae, the combination of fuel molecules such as TAGs is at the expenditure of growth, thus, circumstances need to be improved to optimize TAG production (Scott et al. 2010).

3.1.3 Harvesting and Extraction

As shown in Fig. 2, producing biodiesel requires that biomass be harvested and processed. The obstacles to making biodiesel, in terms of selecting the best method to release fat from the cellular wall, are characterized by the low-energy requirements and economic avoidance of the unreasonable use of solvents such as hexane as well as the increased output of liquid carbon. The safest method is to extract oil without contaminating other cell components such as chlorophyll or DNA. Some of the methods use selective enzymes and decomposition of the cell wall (Scott et al. 2010) (Fig. 3).

3.1.4 The Final Process and Use of Its by-Products

The standard industry method is to extract biodiesel from converted substitutes of TAG transesterification using methanol to obtain methyl esters of fatty acids. Some evidence exists that the fatty acid composition of some types of TAG will be greater in unsaturated acids than is permissible in biofuel components. These substances can then be used to produce glycerol-based products (Scott et al. 2010).

There are a number of methods that can be applied to extract oil from algae, like mechanical compressing, hexane solvent extraction, and so on.

Solvent Extraction

Addition solvent from 0.5% to 0.7% to raw materials for all residuals and leaves for improving the oil extraction. The solvent extraction method can obtain materials with low oil contents. The solvents can use to pre-pressed the high-content materials

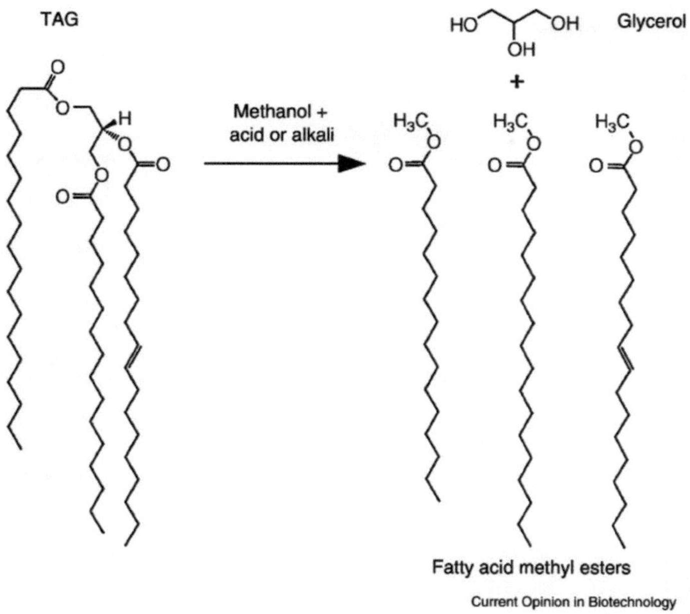

Fig. 3 Esterification of triacylglycerides extracted from algal oil for biodiesel (fatty acid methyl ester) production (Scott et al. 2010)

as oil cakes. This process obtains a high percentage of lipid. Extracting oils or lipids from algae is completed in the following manner. The first step is to obtain algae from a system of open ponds. The algae is then dried in air. Thereafter, the algae is ground. The dried algae samples are then placed in a thimble within a Soxhlet extractor which is placed into a flask filled with extraction solvent and a condenser. Next the solvent is heated until it flows back. The hexane solvent's vapors travel through a distillation arm and flood the chamber housing the thimble. The condenser secures any dissolved vapor in the chamber containing the solid material. The warm solvent floods slowly into the chamber containing the solid material. The desired compound will then dissolve in warm hexane. When the chamber in the Soxhlet extractor is filled with solvent, it is automatically emptied by a siphon arm, with the hexane returning back to the distillation flask (Fig. 4) (Topare et al. 2011).

Oil Extraction Using an Expeller

The expeller method presses the algae mechanically. Algae is acquired from open pond systems and then dried in natural air. When using the expeller method temperatures may exceed 49 °C between the pressed raw material and screw. The screw presses the oil seeds into the cavity of a barrel. Algae is introduced into the expeller on one side of the screw with the oil being output on the other side of the screw. The

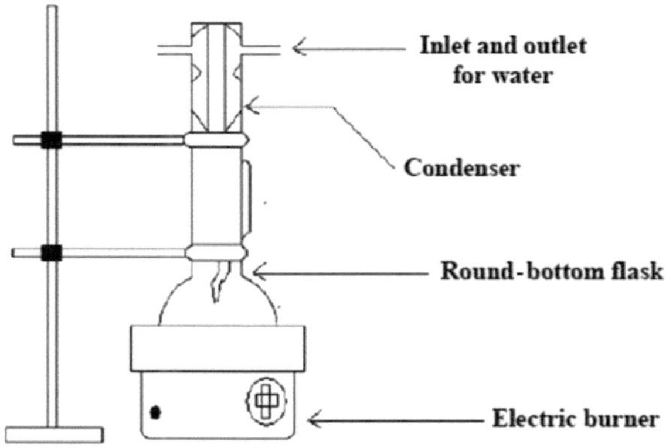

Fig. 4 Solvent method for oil extraction from algae (using a Soxhlet extractor) (Topare et al. 2011)

continuous pressure and friction of the screw drive thereby presses the filamentous algae (Topare et al. 2011).

4 Bioethanol

Bioethanol fuel is an alcohol produced by fermentation, generally using carbohydrate products which are found in sugar and starch crops like corn, sugar cane, and sweet sorghum. Ethanol can be utilized as a fuel for transportation in pure form, normally used with additives to increase the gasoline octane content and improve vehicle emissions (Hossain et al. 2015). During the first generation of bioethanol use, there were concerns about mounting food prices and the use of agricultural fields for the production of bioethanol from feedstock. That problem was countered partially by using lingo-cellulosic materials like crop residues or wastes in second-generation feedstocks. The main advantage of second-generation feedstocks over first-generation feedstocks was the reduced use of food materials and the lesser requirement for land. Nevertheless, purification, production, and several pretreatment requirements have made their production very challenging and uneconomical (John et al. 2011).

Algae used in third-generation biomass used to produce biofuels represents an alternative to first-generation and second-generation biomass because of its high productivity and the ease with which it can be planted (Daroch et al. 2013). The production of ethanol from algae depends on the fermentation of algal polysaccharides, that is, cellulose, starch, and sugar. Under special conditions, the carbohydrate content in microalgae is rich—about 70% (Branyikova et al. 2011). The cell walls of microalgae are split into inner and outer cell wall layers. The cell structure of the outer cell wall can be formed as a trilaminar outer cell wall layer and a thin outer monolayer (Yamada and Sakaguchi 1982). The microalgae outer cell walls contain polysaccharides like agar, alginate, and pectin. Nevertheless, their cell structure can differ from species to species (Yamada and Sakaguchi 1982). Conversely, the inner cell walls of algae are mostly composed of cellulose, hemicellulose, and other substances (Yamada and Sakaguchi 1982). Microalgae is considered a feedstock for producing bioethanol, this is because it has both starch and cellulose in its cell walls (Brennan and Owende 2010). Generally, the polysaccharides and their cellular walls can be fermented to produce ethanol (Hall and Payne 1997).

Their are many ways used to produce ethanol—digestive enzymes (discharging sugars from stored starch), carbohydrate fermentation, distillation, and drying. Ethanol can be used as an alternate to gasoline in gasoline engines by mixing gasoline with ethanol to any percentage. Most current automotive gasoline engines can operate on a bioethanol mix of 15% gasoline or petroleum. Ethanol contains less energy than gasoline. This means it requires much more fuel to produce a similar quantity of energy (Bruhn et al. 2011). Ethanol's advantage is its higher octane ratio, allowing increased engine pressure to increase thermal efficiency. Compared to gasoline, ethanol contains approximately one third of the energy content per unit volume (Hossain et al. 2015).

4.1 Pretreatment of Lingo-Cellulosic Materials

The purpose of pretreatment is to separate or deposit hemicellulose and lignin, decrease cellulose crystallization, and increase material porosity. Pretreatment must fulfill the following requirements: (1) use enzymatic hydrolysis to form sugars; (2) circumvent degradation or carbohydrate loss; (3) avert inhibitory formation of some by-products produced during the hydrolysis and fermentation process; and (4) be economical (Sun and Cheng 2002).

4.2 Bioethanol Production

Extraction of ethanol from biomass is achieved in two stages: carbohydrate hydrolysis to simple sugars (xylose and glucose) and sugar fermentation into alcohol. During the hydrolysis process, the carbohydrate is divided into glucose molecules, where the efficiency of cellulose conversion is based on mechanical and chemical preprocessing (Demirbas 2007). As shown in Fig. 5, the enzymatic hydrolysis process is followed by carbohydrate enzymatic hydrolysis with the assistance of acidic and cellulose enzymes. A high concentration of ethanol can be obtained by the fermentation of C_5 and C_6 sugars and subsequent ethanol distillation (Demirbas et al. 2011).

4.2.1 Ethanol Production from Microalgae

Some species of microalgae are ideal for producing bioethanol using their carbohydrates, which can be extracted, to make fermented sugars. These species of microal-

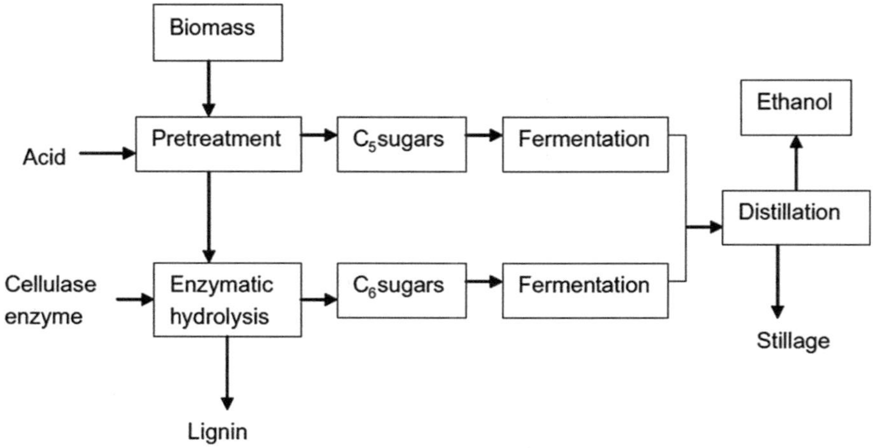

Fig. 5 Process of enzymatic hydrolysis (Demirbas et al. 2011)

gae can produce high levels of carbohydrates rather than lipids—like backup polymers. One such type of algae of importance is blue–green algae, including *Spirogyra* and *Chlorococum* sp. which contain accumulated polysaccharides at high levels in their complex cell walls or as a starch. Such a starch accumulation can be used to produce bioethanol—at high concentrations the blue–green algae of the *Chlorococum* sp. can produce ethanol which is 60% rich. It can be produced from samples that are extracted before the proportion of lipids versus those that are still dry whole cells (Harun et al. 2010; Eshaq et al. 2011).

Microalgae in particular are a potential feedstock for producing bioethanol since they can accumulate starch at levels of about 37% (Hirano et al. 1997). Microalgae used to produce bioethanol through drying is grown in an appropriate aquatic environment. Subsequently the algae is milled and its hydrolyzed mass fermented and distilled (Demirbas and Demirbas 2010). The first step in producing ethanol using microalgae is the use of a mechanical machine or enzyme to release the microalgal starch from the cells. When the cells start to degrade, the fermentation of biomass is initiated by the addition of yeast such as *Saccharomyces cerevisiae* (Nguyen and Vu 2012). The output product of fermentation is ethanol. The mechanical process uses the ethanol is discharged from the cistern and pumped into a chamber to feed the distillation unit. Ethanol is produced by microalgal photosynthesis and intracellular anaerobic fermentation (Pimentel and Patzek 2005; Demirbas 2011). Crop residues of 1 ha can produce 10–100 times more than any other source of oil crop. The cycle of oil crops takes between a few months to 2 or 3 years for full production. However, algae can begin to produce harvestable oil after 3–5 days (Nguyen and Vu 2012).

4.2.2 Ethanol Production from Macroalgae

Seaweed is categorized into three types: red, green, and brown. These types contain different glucans, many polysaccharides made up of glucose, which are considered to be biomaterials with great potential. These seaweeds have low concentrations of lignin (Yanagisawa et al. 2011).

4.3 Energy Extraction from Macroalgal Biomass

4.3.1 Energy Plucking-Out Methods Needed for Dry Macroalgae

Direct Combustion

Direct combustion is the oldest major technique used to obtain energy from dry biomass. This method can provide heat and/or steam, which is used domestically and industrially or for electricity production. High moisture content in the biomass can decrease the heat energy released when compared to the heat generated using dry biomass (20% MC) (Demirbas 2001). The direct combustion of biomass is "possible" only for biomass which has a moisture content of less than 50% (Varfolomeev and Wasserman 2011).

Pyrolysis

Pyrolysis is an alternative thermolytic technique used to convert biomass into fuel. This can be broadly defined as the thermal decomposition process of biomass by heating without interference from air during processing (McKendry 2002; Saidur et al. 2011; Li et al. 2013). Pyrolysis processes are categorized by their temperatures and length of process time, that is, slow, fast, and flash (Ghasemi al. 2012; Li et al. 2013). The pyrolysis process is characterized by long dwelling times at very low heating rates and low reactor temperatures (Milledge et al. 2014), along with the production of char rather than fuel products, whether gaseous or liquid (Brennan and Owende 2010; Ghasemi et al. 2012). Slow or fast pyrolysis shelters are ranging from modern technologies that work at temperatures greater than 500 °C and slow process of vapor retention times for a few seconds or less (Brennan and Owende 2010; Li et al. 2013).

Pyrolysis can produce large amounts of fuel materials relative to the amount of biomass used. The process of pyrolysis can be improved in favor of bio-oil production (a liquid product whose structure depends on the feedstock and pyrolysis procedure used), syngas, or solid char (Ghasemi et al. 2012) depending on the product phase required.

Gasification

The process of gasification is the transformation of organic matter to a combustible gas mixture (syngas) at a high temperature (800–1000 °C) by partial oxidation (Demirbas 2001; Saidur et al. 2011). This process includes the following stages. First the paralysis occurs in a response producing char. Thereafter, the process of gasification occurs, in the presence of a gasifying agent like O_2 or H_2O, which produces syngas. Importantly, the syngas produced by char gasification is much higher than that produced by conventional pyrolysis (Ahmed and Gupta 2010). The gas has a calorific value of about 4–6 MJ m^{-3} (McKendry 2002), and is comprised of a mixture of gases like carbon monoxide (20–30%), hydrogen (30–40%), ethylene (1%), methane (10–15%), nitrogen, and water vapor (Saidur et al. 2011). Syngas can be burned to produce heat or be converted to electricity (Demirbas 2001; McKendry 2002). The syngas produced from gasification can be used to produce methanol and hydrogen forms of fuel (Saidur et al. 2011).

4.4 Energy Plucking-Out Methods Needed for Wet Macroalgae

4.4.1 Hydrothermal Treatments

The process of liquefaction occurs at high pressures and low temperatures. Biomass in a hydrothermal process is converted to hydrocarbon fuel in the presence of hydrogen and a catalyst (McKendry 2002). It is then fermented to produce bioethanol (via anaerobic digestion). The hydrothermal process is considered a pressurized aqueous pyrolysis process (Marcilla et al. 2013). It produces a biofuel by using lower oxygen and moisture contents than the pyrolysis process (Neveux et al. 2014).

Biofuel productivity from hydrothermal liquefaction of microalgae reaches 41% (as percentage mass of original dry microalgae biomass) for *Spirulina* microalgae (Jena and Das 2011), about 45% for *Scenedesmus* microalgae (Vardon et al. 2012), 37% for *Dunaliella* (Minowa et al. 1995), 56% for *Enteromorpha prolifera* sp. (Zhou et al. 2010), 63% for *Laminaria saccharina* (Anastasakis and Ross 2011), and over 49% for *Desmodesmus* (Alba et al. 2012).

4.4.2 Macroalgal Anaerobic Digestion

Seaweeds are comprised of mostly biomass which is suitable for anaerobic digestion (AD) (Sutherland and Varela 2014). In fact, in the 19th century, biogas from algae was used as a source of lighting in an iodine production factory (Milledge et al. 2014). Recently, Tokyo Gas stated that about 20 m^3 of methane gas could be generated from 1 ton of seaweed, and when mixed with natural gas could produce 9.8 kW of

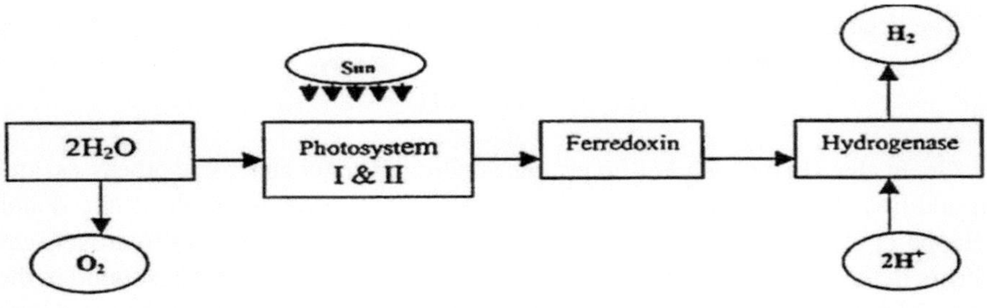

Fig. 6 Mechanism of biophotolysis (Shaishav et al. 2013)

power (Huesemann et al. 2010). The environmental importance of producing biogas from seaweed is based on its ability to decrease greenhouse gas emissions to 42% as compared to the 82% from natural gas (Milledge et al. 2014). In addition, digestion (the material left after the anaerobic process) contains compounds which contain nitrogen and phosphorus, making it a potential fertilizer for seaweed and biological feedstock, thereby providing additional income streams to seaweed anaerobic processing (Milledge et al. 2014).

5 Biohydrogen

The diversity of biofuel sources is very important in energy production (Saifuddin and Priatharsini 2016). Biohydrogen carries the promise of being a clean fuel for use in the future due to increased pollution from fossil fuels and the continued decline in the availability of fossil fuel quantities. Biohydrogen produced from algae is an alternative to the depleting sources of gasoline and is also a clean source of energy (Shaishav et al. 2013).

5.1 Various Processes Used for the Production of Hydrogen from Algae

5.1.1 Direct Biophotolysis

The separation of water molecules under sunlight in the presence of microalgae forms the basis of direct biophotolysis. Microalgae have genetic, metabolic, electron, and enzymatic transport mechanisms in order to produce hydrogen gas. Oxygen and hydrogen is produced by converting a readily available substrate, water, through biophotolysis using solar energy as illustrated in Fig. 6 (Shaishav et al. 2013).

The general reaction for biophotolysis is described by:

$$2H_2O + light = 2H_2 + O_2$$

Green algae like *C. reinhardtii* produce hydrogen under anaerobic processes, and in addition to hydrogen production, the hydrogen is used as an electronic donor (Happe et al. 1994). The conversion of hydrogen-generated ions to hydrogen gas occurs in the medium of electrons by the enzyme hydrogenase which is found in the cells. The energy of light which is absorbed by photosystem I is used to generate electrons that are transferred to ferredoxin by photosynthesis II (Shaishav et al. 2013).

5.1.2 Indirect Biophotolysis

Indirect biophotolysis processes produce hydrogen through blue–green algae like cyanobacteria. The following reactions demonstrate hydrogen formation from water by cyanobacteria (Pinto et al. 2002):

$$12H_2O + 6CO_2 + lightenergy = C_6H_{12}O_6 + 6O_2$$

$$C_6H_{12}O_6 + 12H_2O + lightenergy = 12H_2 + 6CO_2$$

Photosystem II works on the extraction of electrons from water molecules using the energy of sunlight during the process of photosynthesis. The resulting electrons are transferred from water oxidation to Fe-S proteins in ferredoxin on the reduced side of photosystem I. Hydrogenase in the algal stroma allows the electrons to reduce ferredoxin and donate it to two protons to produce a single hydrogen molecule (Shaishav et al. 2013). For the production of photosynthetic hydrogen, cyanobacteria have been identified as perfect candidates. They can produce, in the presence of N_2 and CO_2 in the air, water and mineral salts followed by incubation under argon light and a CO_2 atmosphere that is rapidly becoming a source of energy (Pinto et al. 2002).

5.1.3 Dark Fermentation

Dark fermentation means producing hydrogen in a dark environment in the absence of oxygen, sunlight, and water. Fermentative microorganisms hydrolyze complex organic polymers into monomers that are transformed into a combination of low-molecular-weight organic acids and alcohol by producing bacteria (Schara et al. 2008; Das and Veziroglu 2008). The features of dark fermentation for hydrogen production are: a lack of light; use of different sources of carbon; and production of by-product acids like lactic, butyric, and acetic. However, the disadvantages are: producing a gas mixture which contains carbon dioxide; having the ability to be separated; and having relatively lower hydrogen yields (Saifuddin and Priatharsini 2016).

5.1.4 Photo-Fermentation

Sunlight as an energy source is used for fermentative conversion of organic substrates to H_2 and CO_2 (Sharma and Arya 2017). The reaction of photo-fermentation is demonstrated in the following equation:

$$CH_3COOH + 2H_2O + light = 4H_2 + 2CO_2$$

Purple non-sulfur (PNS) bacteria are used to produce electrons, protons, and CO_2 using sunlight, whereas oxidization of the organic acid substrates occurs by utilization of the tricarboxylic acid cycle (Akkerman et al. 2002; Manish and Banerjee 2008). The advantages of this method include producing organic acids by dark fermentation, removal of environmental pollutants, and use of industrial waste. The disadvantages include the pretreatment of industrial effluent due to the fact that it may be toxic and the need for nitrogen-limited conditions (Mathews and Wang 2009).

6 Conclusions

Biofuel can be obtained from various raw materials—provided it contains cellulose and lignins—as oils, algae, grass, and wood. Microalgae are considered one of the best raw materials for biofuel production. Microalgae such as *C. reinhardtii, D. salina*, and various *Chlorella* species can be used to extract biodiesel; *Chlorococum* species can be used to extract bioethanol; and finally *C. reinhardtii* can be used to extract biohydrogen. The common methods used to extract biofuel are direct combustion, pyrolysis, gasification, and fermentation. The perfect extraction, that is, the easiest, safest, and fastest method is a biochemical technique (fermentation and pyrolysis), which includes photo-fermentation or pyrolyzation of a biomass in the presence of an inert gas.

References

Abdelaziz AEM, Leite GB, Hallenbeck PC (2013) Addressing the challenges for sustainable production of alga biofuels: II. Harvesting and conversion to biofuels. Environ Technol 34:1807–1836

Ahmed II, Gupta AK (2010) Pyrolysis and gasification of food waste: syngas characteristics and char gasification kinetics. Appl Energy 87:101–108

Akkerman I, Janssen M, Rocha J, Wijffels RH (2002) Photobiological hydrogen production: photochemical efficiency and bioreactor design. Int J Hydrogen Energy 27:1195–1208

Alba LG, Torri C, Samori C, Van der Spek J, Fabbri D, Kersten SRA, Brilman DWF (2012) Hydrothermal treatment of microalgae: evaluation of the process as conversion method in an algae biorefinery concept. Energy Fuels 26:642–657

Anastasakis K, Ross AB (2011) Hydrothermal liquefaction of the brown macro-alga *Luminaria saccharina*: effect of reaction conditions on product distribution and composition. Bioresour Technol 102:4876–4883

Branyikova I, Marsalkova B, Doucha J, Branyik T, Bisova K, Zachleder V, Vitova M (2011) Microalgae—novel highly efficient starch producers. Biotechnol Bioeng 108:766–776
Brennan L, Owende P (2010) Biofuels from microalgae—a review of technologies for production, processing, and extractions of biofuels and co-products. Renew Sust Energ Rev 14:557–577
Bruhn A, Dahl J, Nielsen HB, Nikolaisen L, Rasmussen MB, Markager S, Olesen B, Arias C, Jensen PD (2011) Bioenergy potential of UlvaLactuce: bio-mass yield, methane production and combustion. Bioresour Technology 102:2595–2604
Chisti Y (2007) Biodiesel from microalgae. Biotechnol Adv 25(3):294–306
Daroch M, Geng S, Wang G (2013) Recent advances in liquid biofuel production from algae feedstocks. Appl Energy 102:1371–1381
Das D, Veziroglu TN (2008) Advances in biological hydrogen production processes. Int J Hydrogen Energy 33:6046–6057
Demirbas A (2001) Biomass resource facilities and biomass conversion processing for fuels and chemicals. Energy Convers Manag 42:1357–1378
Demirbas A (2007) Progress and recent trends in biofuels. Prog Energy Combust Sci 33:1–18
Demirbas MF (2011) Biofuels from algae for sustainable development. Appl Energy 88:3473–3480
Demirbas A, Demirbas MF (2010) Algae energy: algae as a new source of biodiesel. Springer Science & Business Media, Berlin
Demirbas MF, Balat M, Balat H (2011) Biowastes-to-biofuels. Energy Convers Manag 52(4):1815–1828
Eshaq FS, Ali MN, Mohd MK (2011) Production of bioethanol from next generation feed-stock alga *Spirogyra* species. Int J. Eng. Sci. Technol. 3:1749–1755
Ghasemi Y, Rasoul-Amini S, Naseri AT, Montazeri-Najafabady N, Mobasher MA, Dabbagh F (2012) Microalgae biofuel potentials (Review). Appl Biochem Microbiol 48:126–144
Hall J, Payne G (1997) Factors controlling the growth of field population of *Hydrodictyon reticulatum* in New Zealand. J Appl Phycol 9:229–236
Happe T, Mosler B, Naber JD (1994) Induction, localization and metal content of hydrogenase in the green alga *Chlamydomonas reinhardtii*. Eur J Biochem 222:769–774
Harun R, Danquah MK, Forde GM (2010) Microalga biomass as a fermentation feedstock for bioethanol production. J Chem Technol Biotechnol 85:199–203
Hirano A, Ueda R, Hirayama S (1997) CO_2 fixation and ethanol production with microalga photosynthesis and intracellular anaerobic fermentation. Energy 22:137–142
Hossain S, Salleh A (2008) Biodiesel fuel production from algae as renewable energy. Am J Biochem Biotechnol 4(3):250–254
Hossain NB, Basu JK, Mamun M (2015) The production of ethanol from microalgae Spirulina. Procedia Eng 106:733–738
Huesemann M, Roesjadi G, Benemann J, Metting FB (2010) Biofuels from microalgae and seaweeds. In: Biomass to biofuels. Blackwell Publishing Ltd.: Oxford, UK, pp. 165–184
Jena U, Das KC (2011) Comparative evaluation of thermochemical liquefaction and pyrolysis for bio-oil production from microalgae. Energy Fuels 25:5472–5482
John RP, Anisha GS, Nampoothiri KM, Pandey A (2011) Micro and macroalga biomass: a renewable source for bioethanol. Bioresour Technol 102(1):186–193
Lang X, Dalai AK, Bakhshi NN, Reaney MJ, Hertz PB (2002) Preparation and characterization of biodiesels from various bio-oils. Bioresour Technol 80:53–62
Li L, Rowbotham JS, Greenwell CH, Dyer PW (2013) An introduction to pyrolysis and catalytic pyrolysis: versatile techniques for biomass conversion. In: Suib SL, Elsevier Ed (eds) New and future developments in catalysis: catalytic biomass conversion. Amsterdam, The Netherlands, pp 173–208
Liau BC, Shen CT, Liang FP, Hong SE, Hsu SL, Jong TT, Chang CM (2010) Supercritical fluids extraction and anti-solvent purification of carotenoids from microalgae and associated bioactivity. J Supercrit Fluids 55:169–175
Manish S, Banerjee R (2008) Comparison of biohydrogen production processes. Int J Hydrogen Energy 33(1):279–286

Marcilla A, Catalá L, García-Quesada JC, Valdés FJ, Hernández MR (2013) A review of thermochemical conversion of microalgae. Renew Sustain Energy Rev 27:11–19

Mata TM, Martins AA, Caetano NS (2010) Microalgae for biodiesel production and other applications. Renew Sustain Energy Rev 14:217–232

Mathews J, Wang G (2009) Metabolic pathway engineering for enhanced biohydrogen production. Int J Hydrogen Energy 34:7404–7416

McKendry P (2002) Energy production from biomass (part 3): Gasification technologies. Bioresour Technol 83:55–63

Medipally SR, Yusoff FM, Banerjee S, Shariff M (2015) Microalgae as sustainable renewable energy feedstock for biofuel production. Bio Med Res Int 2015:519–513

Metzger P, Largeau C (2005) *Botryococcus braunii*: a rich source for hydrocarbons and related ether lipids. Appl Microbiol Technol 66:486–496

Milledge JJ, Smith B, Dyer PW, Harvey P (2014) Macroalgae—derived biofuel: a review of methods of energy extraction from seaweed biomass. Energies 7:7194–7222

Minowa T, Yokoyama S, Kishimoto M, Okakura T (1995) Oil production from alga cells of *Dunaliella tertiolecta* by direct thermochemical liquefaction. Fuel 74:1735–1738

Neveux N, Yuen AKL, Jazrawi C, Magnusson M, Haynes BS, Masters AF, Montoya A, Paul NA, Maschmeyer T, De Nys R (2014) Biocrude yield and productivity from the hydrothermal liquefaction of marine and freshwater green macroalgae. Bioresour Technol 155:334–341

Nguyen THM, Vu VH (2012) Bioethanol production from marine algae biomass: prospect and troubles. J Viet Environ 3(1):25–29

Nigam PS, Singh A (2011) Production of liquid biofuels from renewable resources. Prog Energy Combust Sci 37:52–68

Pimentel D, Patzek TW (2005) Ethanol production using corn, switchgrass, and wood; biodiesel production using soybean and sunflower. Natural Resour Res 14:65–76

Pinto FAL, Troshina O, Lindblad P (2002) A brief look at three decades of research on cyanobacterial hydrogen evolution. Int J Hyd Ener 27:1257–1264

Rösch C, Skarka J, Wegerer N (2012) Materials flow modeling of nutrient recycling in biodiesel production from microalgae. Bioresour Technol 107:191–199

Saidur R, Abdelaziz EA, Demirbas A, Hossain MS, Mekhilef S (2011) A review on biomass as a fuel for boilers. Renew Sustain Energy Rev 15:2262–2289

Saifuddin N, Priatharsini P (2016) Developments in bio-hydrogen production from algae: a review. Res J Appl Sci Eng Technol 12(9):968–982

Schara V, Maeda GT, Wood TK (2008) Metabolically engineered bacteria for producing hydrogen via fermentation. Microb Biotechnol 1(2):107–125

Scott SA, Davey MP, Dennis JS, Horst I, Howe CJ, Lea-Smith DJ, Smith AG (2010) Biodiesel from algae: challenges and prospects. Curr Opin Biotechnol 21:277–286

Shaishav S, Singh RN, Satyendra T (2013) Biohydrogen from Algae: Fuel of the Future. Int Res J Environ Sci 2(4):44–47

Sharif ABMH, Nasrulhaq AB, Majid HAM, Chandran S, Zuliana R (2007) Biodiesel production from waste cooking oil as environmental benefits and recycling process. A review. Asia Biofuel Conference Book. Dec 11–13, Singapore

Sharma A, Arya SK (2017) Hydrogen from alga biomass: a review of production process. Biotechnol Rep 15:63–69

Shuba ES, Kifle D (2018) Microalgae to biofuels: 'Promising' alternative and renewable energy. Review. Renew Sust Energy Rev 81:743–755

Sijtsma L, Swaaf ME (2004) Biotechnological production and applications of the w-3-polyunsaturated fatty acid docosahexaenoic acid. Appl Microbiol Biotechnol 64:146–153

Spolaore P, Joannis-Cassan C, Duran E, Isambert A (2006) Commercial applications of microalgae. J Biosci Bioeng 101:87–96

Sun Y, Cheng J (2002) Hydrolysis of lignocellulosic materials for ethanol production: a review. Bioresour Technol 83:1–11

Sutherland A, Varela J (2014) Comparison of various microbial inocula for the efficient anaerobic digestion of Laminaria hyperborea. i, 14. https://doi.org/10.1186/1472-6750-14-7
Topare NS, Rauta SJ, Renge VC, Khedkar SV, Chavan YP, Bhagat SL (2011) Extraction of oil from algae by solvent extraction and oil expeller method. Int J Chem Sci 9(4):1746–1750
Vardon DR, Sharma BK, Blazina GV, Rajagopalan K, Strathmann TJ (2012) Thermochemical conversion of raw and defatted alga biomass via hydrothermal liquefaction and slow pyrolysis. Bioresour Technol 109:178–187
Varfolomeev SD, Wasserman LA (2011) Microalgae as source of biofuel, food, fodder, and medicines. Appl Biochem Microbiol 47:789–807
Voloshin RA, Rodionova MV, Zharmukhamedov SK, Veziroglu TN, Allakhverdiev SI (2016) Review: biofuel production from plant and algae biomass. Int J Hydrogen Energy 41:17257–17273
Wang B, Li Y, Wu N, Lan CQ (2008) CO_2 bio-mitigation using microalgae. Appl Microbiol Biotechnol 79:707–718
Yamada T, Sakaguchi K (1982) Comparative studies on Chlorella cell walls—induction of protoplast formation. Arch Microbiol 132:10–13
Yanagisawa M, Nakamura K, Ariga O, Nakasaki K (2011) Production of high concentrations of bioethanol from seaweeds that contain easily hydrolysable polysaccharides. Process Biochem 46(11):2111–2116
Zhou D, Zhang L, Zhang S, Fu H, Chen J (2010) Hydrothermal liquefaction of macroalgae Enteromorpha prolifera to bio-oil. Energy Fuels 24:4054–4061

Strategies to Improve Enzymes via Solid-State Fermentation

Indu Bhushan, Manjot Kour and Guneet Kour

Abstract Solid-state fermentation (SSF) is the fermentation process which occurs on a solid surface in the absence of "free" water, where the moisture is absorbed to the solid matrix. SSF is gaining an advantageous edge over other fermentation techniques due to its less complexity and more proximity to the normal environment of many microorganisms. On the other hand, a difficulty arises while estimating the biomass concentration in solid-state fermentation. Various factors like direct product application, the increased concentration of the product, less cost of production, and reduced energy requirement are responsible in making SSF as one of the potent technologies for various enzyme productions as seen in case of cellulase, tannase, and lipase. Improvisation of cellulase production in solid-state fermentation can be achieved to a greater extent by making use of varying degrees of substrates which are lignocellulosic in nature, the implicated microorganisms, culture, and process parameters like moisture content and water activity, nutrients diffusion, size of substrate particle, pH, temperature, surfactants, and bioreactor designs. Submerged fermentation whereas holds a different place in terms of various types of fermentations as it has only one major problem related to the oxygen transfer to microorganisms which in turn depends on the configuration, size, and the agitation/aeration system used in the reactor. In order to characterize oxygen transfer, a parameter is known as K La (oxygen transfer coefficient) whose value gives the estimation that how much of the oxygen is transferred by the equipment independent of the reactor volume and hence, for scale-up studies, it becomes an important parameter. In case of antioxidants production using SSF, it was observed that pomace tends to increase the antioxidant activity convergent with an increase in activity of β-glucosidase. Different studies tend to show that *P. floridensis* as an important organism used during the production of lingo cellulolytic enzyme and consecutive advancement in in vitro digestibility of wheat straw has been carried out to a larger extent.

Keywords Fermentation · Biomass cellulases · Lipases · Tannase · Phenolic oxidants

I. Bhushan (✉) · M. Kour · G. Kour
Shri Mata Vaishno Devi University, Katra, India
e-mail: sharmasmvdu92@gmail.com

N. Srivastava et al. (eds.), *Sustainable Approaches for Biofuels Production Technologies*, Biofuel and Biorefinery Technologies 7,
https://doi.org/10.1007/978-3-319-94797-6_7

1 Introduction

The fermentation process which takes place in such a situation where the free water is nearly or completely absent is called solid-state fermentation (SSF). These days definition of solid-state fermentation is taking a new turn as according to number of citations it has been defined as a process where we find microorganisms growing on damp particles of solid materials embedded in the form of beds, wherein there is a continuous flowing gas phase in the spaces between the particles (Behera and Ray 2016). In order to produce various chemicals used in industries, pharmaceutical products, feed, fuel, etc., SSF has come up as one of the prospective technologies. In SSF processes, natural raw materials are generally employed as carbon and energy source. Solid matrix in SSF is composed of chemically inactive material which requires a solution rich in nutrients. The solid material importantly should consist of sufficient amount of water in the form of moisture. Nature of the substrate has an indirect relation with the achievement of higher biochemical process rate because if the content of water absorbed is more than the required compared to dry weight of solid matrix, then it will result in high water activity (a_w) on the interface, leading to biochemical process increase (Mitchell 2011). In case of lower water activity, there is decrease in dispersion of nutrients through the solid matrix, whereas in case of higher water activity compaction in substrate particles takes place. Therefore, appropriate water activity and appropriate level of moisture in the solid substrate forms the essential elements for SSF processes. Along with other factors, surface area of solid substrate plays an important role. It should be generally large, in the range of 10^3 to 10^6 m2/cm^3 so that there is optimum growth on the interface (Manan and Webb 2018). Small size of substrate particles offers larger surface area for microbes to attack but at the same time pose hindrance in respiration and aeration due to interparticle space availability constraint. For the reason of cost-effectiveness in bioprocess optimization, sometimes it becomes necessary to compromise with the size of particles to be used in the process (Durand 2003), for instance, wheat bran, one of the substrates which is frequently used in fine and coarse forms for SSF. In order to achieve maximum production, most of SSF processes make use of mixture of these two forms. Substrates which are solid in nature provide a favorable environment to the bacteria and fungi. The characteristic hyphal growth pattern of filamentous fungi on the superficial surface of the substrate particles makes them as the best studied candidate for SSF. Different types of agriculture-based crop like barley and agro-industrial based leftover of rice and wheat bran, different types of oil cakes of coconut oil, palm kernel, soybean, sugarcane bagasse, cassava bagasse, fruit pulps, seeds like tamarind seeds, jackfruit seeds, corn cobs, etc. are the substrates which are frequently used for SSF processes (Mienda and Idi 2011). While growing on these substrates, microorganisms secrete various hydrolytic exoenzymes which help in breakdown of some complex carbon sources and nutrients which in turn promote biosynthesis and other microbial activities.

With the advent of technology and understanding of certain domains of biochemical engineering, more precisely mathematical modeling and fermenter design, it has become possible to scale up various SSFs. Few fermenter designs and mathematical models have been adopted for commercial purposes. If these trends keep on flourishing at such a pace, a time will come when SSF technology would be developed quite well and would come in shoulder to shoulder with submerged fermentation technology (Viniegra-González et al. 2003).

There are numerous attractive advantages of SSF including the production of extracellular enzymes that are stable at various temperatures and pH ranges and high production volumes which is nearly 5.6 times larger than submerged fermentation. SSF is commonly used for the enzyme production because it encompasses the production of extremely concentrated crude enzymes that are associated with low costs for extraction and purification (Muthusamy and Ps 2013).

2 Improvement Strategies

2.1 Cellulase Production

Cellulases are enzymes that catalyze hydrolysis of α-1,4-D-glucan bonds in cellulose resulting in the formation of simpler products such as glucose, cellobiose, and cello-oligosaccharides. According to sequence analysis, from 82 families classified as glycoside hydrolase, 13 were identified as cellulases. Cellulases fall under the category of commonly studied enzymes such as cellobiohydrolases, glucosidases, and endoglucanases. Endoglucanases have a peculiar nature of producing nicks in the cellulose polymer due to which the reducing and nonreducing ends of the polymer are exposed to the environment. Cellobiohydrolases catalyze both reducing terminals and nonreducing terminals forming cello-oligosaccharides besides cellobiose products. Thereafter, glucosidases hydrolyze cellobiose liberating glucose. The action of cellulase complex consisting of cellobiohydrolases, glucosidases, and endoglucanases is synergistical in nature so that the crystalline cellulose gets converted to glucose. Due to wide range of applications, cellulases are the third largest industrial enzyme worldwide. These enzymes are also used as detergent enzymes and animal feed additives. In case the major transportation fuel was ethanol extracted from lignocellulosic biomass via enzymatic route, it will in turn make cellulase the largest volume industrial enzyme. There are a wide variety of cellulase-producing microorganisms comprising several anaerobic bacteria and fungi such as white-rot and soft-rot fungi. Cellulases derived from filamentous fungi for instant *Fusarium*, *Humicola*, *Penicillium*, *Phanerochaete*, *Trichoderma*, etc. are used for industrial applications since filamentous fungi and aerobic bacteria generally secrete free molecules of cellulases. As compared to yeast or bacteria, filamentous fungi cause difficulties in mass transfer and this is primarily due to its characteristic growth pattern. In order to overcome this problem, efficient technologies have been developed which in turn are leading to effective and high titer production of antibiotics, organic acid, and native enzymes. One of the most important cellulase-producing microorganisms which is studied in detail is *Trichoderma reesei* that produces cellobiohydrolases of two types, CBH I and CBH II along with dual types of endoglucanases which consist of EG1 along with EG2. These enzymes are roughly in the proportion of 60:20:10:10, respectively, that collectively contribute about 90% of the enzymes. On contrary, less than 1% is contributed by seven glucosidases—BGLI to BGLVII (Singhania et al. 2010).

2.2 Cellulase Prerequisite Characteristics for Bioconversion

To achieve optimal biomass conversions, explicit features are required such as better thermotolerance, high enzyme activity, better tolerance to extreme pH, and decreased feedback inhibition. Cellulases which are secreted by various filamentous fungi such as *T. reesei* are acidic in nature. Acidic cellulases are preferred for those bioconversions where acidic pretreatment is given and while working with cocktail of acidic enzymes that require pH optima between 4 and 6. Accelerase®1500 is a trade name of cellulase that has an optimal pH range of 4.6 to 5.0 but below pH 4.0 or above pH 7.0 it becomes inactivated. Usually, celluloses work efficiently at 50 °C temperature and even Accelerase®1500 works efficiently in the temperature range of 50–65 °C. Use of single state fermentation to produce cellulase is quickly gaining attention as a technology which is very cost-effective especially due the use of microorganisms like fungi which produce reasonably large-scale cellulase due to the fermentation conditions that are quite similar to their natural conditions. Chahal had reported that *T. reesei* culture in SSF gives higher yield of cellulases as compared to the cultures in liquid. One of the important consequences of high production titer of cellulases by SSF is that it reduces downstream processing, thus decreasing the operation cost. Apart from various agro-based substrates, various wastes from agricultural domain can also be used as effective substrates for the process of enzyme production under SSF. This has even been validated through a review by Nigam and Singh. Pandey et al. described SSF technology for cellulase production and hence reported that SSF (a future technology) is useful for industrial enzyme production. For the very first time, Dutta et al. analyzed that when cellulase is produced by *Penicillium citrine* following SSF, it shows tolerance to alkali environment. The SSF is considered as a beneficial technology for the production of cellulase in bioconversion as purity is not considered important requirement for this application. SSF is an attractive technology wherein production conditions if optimized will result in better economical production of cellulase. SSF is better compared to SmF as it offers less catabolite repression, better productivity, increased product yield, and less generation of effluent. SSF with improved technology such as better operation control and enhanced bioreactor design can provide promising system for cellulase production (Singhania et al. 2010).

2.3 Tannase Production

Till date, many citations have shown interesting advantages of tannase production with the help of SSF (solid-state fermentation) as compared to submerged. Many scientists have studied the production of enzyme where they have observed solid-state fermentation with the help of various agro-wastes substrate rich in tannins. Leaves of Jamun are optimal source to produce the enzyme using SSF. Throughout SSF, maximum tannase production has been observed at 31.1 °C for incubating at about 96 long hours. However, it has been studied that carbon sources and other nitrogen sources when added to the medium do not affect tannase production. Influence of pH and temperature has been studied widely during the process of tannase activity and during the production of large amount of gallic acid from large amount of tannin-rich agro-waste by Reddy and Rathod. In order to produce tannase by the process of SSF, substrates with large amount of tannin content are used. The substrate is allowed to get completely moistened with large amount of minerals in the form of solution which is then inoculated with the selected organism. Sugarcane bagasse, creosote bush leaves (*Larrea tridentata*), oak galls (*Quercus infectoria*), large amount of sumac leaves (*Rhus coriaria*), myrobalan fruit (*Terminalia chebula*), sorghum leaves (*Sorghum vulgaris*), and Indian gooseberry leaves (*Phyllan thusemblica*) are considered to be the natural supports used to produce tannase on large scale. Studies have shown that the supports like polyurethane foam in combination with nutrient media are used on large scale. Modified solid-state fermentation (MSSF) was used for continuous gallic acid production and tannase production using *R. oryzae* rich strain. MSSF tends to increase tannase and gallic acid production yield by 1.6 and 4 times correspondingly with traditional SSF systems (Muthusamy and Ps 2013).

2.4 Lipase Production

Improvement in lipase production has taken place with the help of modifications that have been implemented in the nutrient source using *Rhizopus homothallicus* which is cultured with the help of process that involves SSF. *R. homothallicus* has been used for lipase production in solid-state fermentation (SSF) with the help of sugarcane bagasse as a support which is then impregnated with an adequate amount of medium containing liquid. It was observed that modification in nutrient present in the media was done for lipase production. Lipase production is largely affected due to nutrients that influence and affect the growth which mainly include urea, olive oil, and huge amount of oligo-elements. Previous studies reveal that improved and better medium provides good results for kinetic studies for growth and lipase production (Ramos-Sánchez 2015). An interdependence is observed to exist between the profiles in the presence of lipase and that of CO_2 production, pH changes, and O_2 consumption during lipase production where an incubation period of 12 h revealed a reading of 827 U/g DM. This production has been observed to show large increase in lipolytic activity in comparison to the results which were obtained using known medium to produce lipase. The results were then analyzed to be promising as this strain tends to produce high concentrations of lipase in an inexpensive and reliable medium, which contributes to its purification. In addition to this, the extraction of lipase from the solid medium was also studied to observe the effect, and hence efficiency in the recovery of the enzyme was attained with the help of Triton X-100 at 0.8% (w/v) (Rodríguez González 2006).

2.5 Phenolic Oxidants Production from Cranberry Pomace

Cranberry processing industry has reported production of by-product like Cranberry pomace which can be used extensively for the production of a large amount of phenolic ingredients that are value added. The process of pomace bioprocessing with the help of solid-state fermentation (SSF) and making use of food grade fungi has provided a distinctive and new strategy to enhance various properties especially those of nutraceutical and to produce a large amount of functional and other ingredients. Many functional phytochemicals occur as glycosides or their derivative forms which have comparatively reduced biological and physical activity to a large extent. Therefore, food grade fungus *Rhizopus oligosporus* has been used widely to develop this strategy (Vattem and Shetty 2002). One of the studies reveals that SSF of cranberry pomace has been carried out continuously for 16 days using oxygen sources, nitrogen sources such as ammonium nitrate (NH_4NO_3), and large amount of hydrolysate rich in fish protein (FPH). Nitrogen and oxygen treatments, however, tend to show an increase in water and phenolics which were extracted by 15–25% by day 10 in cranberry pomace. Also, it has been seen that antioxidant protection factor is maximum on 15th day in case of both nitrogen and oxygen treatments and was observed to be 22–27% advanced than that for water extracts and 16.7–19.7% for extracts of ethanol, respectively. The DPPH radical inhibition (DRI) capacity has been seen to increase by 6% for the NH_4NO_3 supplementation and steadily decreases for FPH treatment extracts with water. However, no variation is observed in case of ethanol extracts (White et al. 2010). Activity of β-glucosidase tends to increase by 65-fold in case of other treatment and by 90-fold in case of FPH treatment and with the increasing amount in phenolics which can be extracted and checked for antioxidant activity. HPLC indicates that ellagic acid tends to increase by 4–8-folds in extracts containing water for both oxygen and nitrogen treatments and therefore differences in diphenyl profiles throughout the SSC are examined with this technique. In case of ethanol extracts, this increase was observed to be between 15 and 25%. Hence, it was observed that pomace tends to increase the antioxidant activity convergent with an increase in activity of β-glucosidase. It has been observed that the ellagic acid was seen in HPLC profile, as a component having enriched anti-carcinogenic properties. Function of antioxidant is, however, observed to show fluctuations for preventing major diseases linked with oxidation such as cancer and CVD. This innovative approach using SSF has been widely used to enhance and increase everyday phytochemicals for the sake of food and feed use to a great extent (Vattem and Shetty 2002).

2.6 Production of Lignocellulolytic Enzymes

It has been observed that increasing wheat straw digestibility degradation by wheat rot fungus has been done which has resulted in its improved value as animal feed. Also, the effect of large amount of moisture content, adequate amount of nitrogen sources inorganic in nature (NH_4Cl) and extracts containing malt sugar on lingo cellulolytic enzymes, and difference in other chemical components and amount of digestibility of wheat straw have been widely observed. Laccase production increases up to 36-fold with a wide increase in moisture content. However, enhancement in the production of CMCase and xylanase to a large extent was significant ($p < 0.05$) which was observed using these supplements. In vitro digestibility has been observed to upscale largely by almost 51% with a loss of 27.5% in lignin and 15.6% in overall organic matter. However, some of the findings tend to show that *P. floridensis* as an important organism used during the production of lingo cellulolytic enzyme and consecutive advancement in in vitro digestibility of wheat straw has been carried out to a larger extent (Sharma and Arora 2010).

2.7 Role of Temperature Control in SSF

It has been known that an important constituent in SSF is temperature control. Previous studies reveal that in continuous mixing, aseptic paddle mixing is done profitably for SSF with *Aspergillus oryzae* on a large amount of wheat kernels. It was observed that mixing continuously improves control in temperature and prevents homogeneities. However, it has been observed that rates of respiration that are observed in this organization can be compared to small and isothermally unmixed beds, showing that stirring continuously did not, however, cause extensive harm to the fungus/kernels entirely. However, it has been observed that increase in scale-up calculations for the paddle mixer is observed to show that cooling in walls becomes insufficient at the 4-m^3 scale for a fungus that grows abruptly like *Aspergillus oryzae*. In contrast to this evaporative cooling, temperature tends to be a very important aspect of systems with large-scale mixed constituents. Some experiments tend to show that addition of water is necessary when evaporative cooling is done to maintain sufficiently excess water activity of the solid matrix used as substrate. The process of mixing is, however, observed to be an important and necessary measure to make sure that addition of water is homogeneous in SSF. Also, process control by automation can be achieved with the help of enthalpy balance. This was validated using paddle mixer through experiments. This has shown that mixing continuously provides promising possibilities for control of moisture, and therefore temperature control in solid-state fermentation becomes an integral factor for various productions (Nagel et al. 2001).

3 Conclusion

There is well-established fact that solid-state fermentation is one of the promising technologies to produce large number of industrially important enzymes. Tannase production initially started in early years of microbiology but still more research is needed in order to unravel many unknown facts about its efficient extraction and commercial production. Antioxidant production with the help of SSF has lead to opening of many paths including the one toward the prevention of some deadly oxidation-related diseases like cancer. SSF is preferred over various other fermentation technologies like SmF due to various reasons. One of them is the requirement of less amount of energy for the oxygen supply so that the system can cope up with the high oxygen demand.

Acknowledgements We would love to thank School of Biotechnology, Shri Mata Vaishno Devi University, Katra, Jammu, J&K.

References

Behera SS, Ray RC (2016) Solid state fermentation for production of microbial cellulases: recent advances and improvement strategies. Int J Biol Macromol 86:656–669
Durand A (2003) Bioreactor designs for solid state fermentation. vol 13, pp 113–125
Manan M, Webb C (2018) Estimation of growth in solid state fermentation: a review. vol 14, pp 61–69
Mienda B, Idi A (2011) Microbiological features of solid state fermentation and its applications. An overview. vol 2, pp 21–26
Mitchell DA et al (2011) 2.25—Bioreactors for solid-state fermentation A2. In Moo-Young M (ed) Comprehensive biotechnology, 2nd edn. Academic Press, Burlington, pp 347–360
Muthusamy C, Ps B (2013) Tannase: source, biocatalytic characteristics, and bioprocesses for production. pp 259–293
Nagel FJ et al (2001) Temperature control in a continuously mixed bioreactor for solid-state fermentation. Biotechnol Bioeng 72(2):219–230
Ramos-Sánchez L et al (2015) Fungal lipase production by solid-state fermentation. vol 5, pp 1–9
Rodríguez González J et al (2006) Improving lipase production by nutrient source modification using *Rhizopus homothallicus* cultured in solid state fermentation. Vol 41, pp 2264–2269
Sharma RK, Arora DS (2010) Production of lignocellulolytic enzymes and enhancement of in vitro digestibility during solid state fermentation of wheat straw by *Phlebia floridensis*. Bioresour Technol 101(23):9248–9253

Singhania RR et al (2010) Advancement and comparative profiles in the production technologies using solid-state and submerged fermentation for microbial cellulases. Enzym Microb Technol 46(7):541–549

Vattem DA, Shetty K (2002) Solid-state production of phenolic antioxidants from cranberry pomace by *Rhizopus oligosporus*. Food Biotechnol 16(3):189–210

Viniegra-González G et al (2003) Advantages of fungal enzyme production in solid state over liquid fermentation systems. Biochem Eng J 13(2):157–167

White BL, Howard LR, Prior RL (2010) Proximate and polyphenolic characterization of cranberry pomace. J Agric Food Chem 58(7):4030–4036

Second Generation Bioethanol Production: The State of Art

Hemansi, Rishi Gupta, Garima Yadav, Gaurav Kumar, Anju Yadav, Jitendra Kumar Saini and Ramesh Chander Kuhad

Abstract Second generation bioethanol from waste lignocellulosic biomass is a sustainable solution to the problems of diminishing petroleum reserves, issues over national security and environmental deterioration due to GHG emissions. The production of second generation bioethanol is a complex process and consist several steps including biomass pretreatment, saccharification of cellulosics followed by microbial fermentation and product recovery. In this chapter, an attempt has been made to review the process steps of bioethanol production from plant biomass and their respective scope of improvement. Afterwards, the global and national status of bioethanol production and various policies governing its commercialization have also been dealt with. The chapter also summarizes the energy balance, mass balance, life cycle analysis studies and techno-economic evaluation of lignocellulosic bioethanol production carried out by various researchers. Moreover, the technological barriers and alternatives investigated to overcome the challenges in second generation bioethanol production process are also discussed.

Keywords Second generation bioethanol · Lignocellulosics · GHG emission
Pretreatment · Fermentation

Hemansi · G. Yadav · A. Yadav · J. K. Saini (✉)
Department of Microbiology, School of Interdisciplinary and Applied Life Sciences, Central University of Haryana, Mahendergarh 123031, Haryana, India
e-mail: me_jk@rediffmail.com; jitendrasaini@gmail.com

R. Gupta
Department of Biotechnology, School of Interdisciplinary and Applied Life Sciences, Central University of Haryana, Mahendergarh 123031, Haryana, India

G. Kumar · R. C. Kuhad
Department of Microbiology, University of Delhi South Campus, New Delhi 110021, Delhi, India

R. C. Kuhad
Central University of Haryana, Mahendergarh 123031, Haryana, India

Present Address
R. Gupta
SVI Research, SVI Analytica, Gurugram 122016, Haryana, India

N. Srivastava et al. (eds.), *Sustainable Approaches for Biofuels Production Technologies*, Biofuel and Biorefinery Technologies 7,
https://doi.org/10.1007/978-3-319-94797-6_8

1 Introduction

Current scenario of declining fossil fuel reserves and soaring petroleum prices, concerns over the national energy security and in particular dependence on oil-import have led researchers all over the world to search for an alternative transportation fuel. Additionally, the global climate change and environmental impacts of fossil fuels have also heightened the awareness of replacing fossil fuels from our present way of life (Charles et al. 2007). According to reports from Environmental Protection Agency (EPA), since 1970, the CO_2 emission has almost doubled and GHG (Green House Gas) emissions have increased by 78% due to rapid industrialization and accelerated consumption of fossil fuels (US EPA 2016). As a consequence, governments and industries globally are putting various measures to provide suitable solutions to offset these problems; however, still petroleum is the chief source which is being used to meet the world's fuel demands.

For transportation sector, unconventional energy carriers like H_2 and electricity have been successfully developed but their large-scale application is marred by their lower energy density and storage-related issues (Agrawal et al. 2007). Therefore, it seems more convenient to use liquid transportation fuel through existing infrastructure. Biofuels in general and bioethanol, in particular, are the most promising clean fuel, which can be easily integrated in the prevailing transportation system. Although ethanol's energy content is roughly 2/3rd of gasoline, it has higher research octane number (107) than gasoline (91–99) (Lynd 1996). Moreover, researchers have shown that ethanol can be used up to 85% (v/v) in vehicles without major modifications (Balat et al. 2008) with associated benefits of being bio-renewable in nature, generation of less harmful emission and therefore, being environmentally sustainable and reduced dependence upon petroleum resources. Burning of petroleum-based fuels generate more harmful discharges when compared to that of ethanol (Wyman and Hinman 1990) and therefore, application of even E10 blend (10% ethanol in gasoline) results in up to 20% decreased GHGs. Further increase in ethanol blending has more prominent effect on reduced emissions of NO_x, SO_x, and particulate matter. Due to the associated benefits of using ethanol as an alternative or supplementary transportation fuel, there has been a global upsurge of interests in research and development of bio-based fuels from renewable biomass-based resources in a sustainable manner.

Currently, almost all of the commercially available bioethanol in United States, Brazil, and the European Union is produced from either starch- or sugar-rich crops, which is referred as 'first generation (1G) bioethanol'. For 1G bioethanol sugars derived from cane, molasses or corn starch are used as primary starting material. Production of bioethanol from such resources is expected to increase further in the coming few years (Goldemberg 2007). However, due to food nature of such resources, competition of the bioethanol fuel with the food is also expected to increase together with the expected deforestation to achieve higher production and further negative environmental impacts (Hahn-Hägerdal et al. 2006; Tenenbaum 2008). Therefore to combat the problems associated with the use of first generation bioethanol, interest

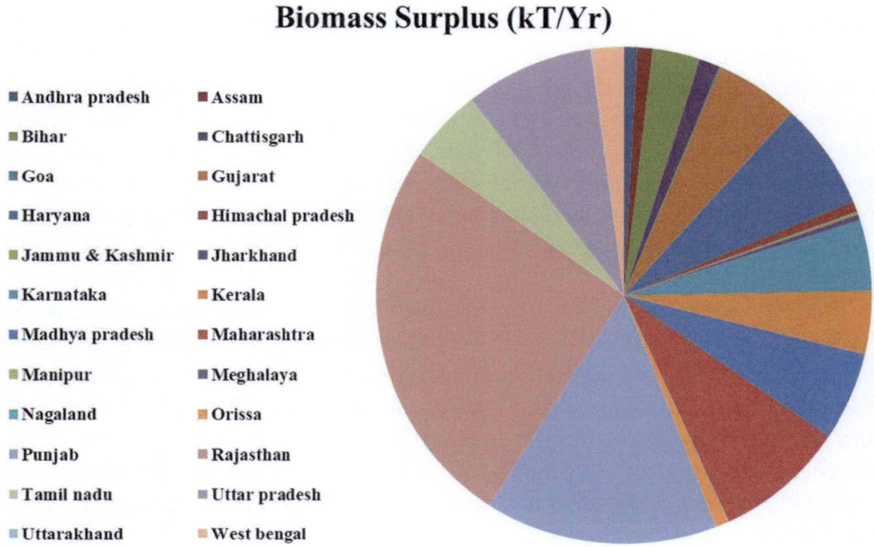

Fig. 1 Availability of lignocellulosic biomass in different states of India

was shifted to generate 'second generation or 2G bioethanol' from lignocellulosic non-food crops (e.g. *Prosopis, Miscanthus*) or waste plant biomass, such as crop wastes, rice and wheat straw, cotton stalk, etc.) or other waste resources like municipal solid wastes (MSWs) (Claassen et al. 1999).

Lignocellulosic biomass is the most promising feedstock considering its great availability, low cost and non-competence with the food demands. The availability of lignocellulosic biomass in Indian context is shown in Fig. 1. The conversion of lignocellulosic biomass to bioethanol is a multi-step process. The structural carbohydrate polymers in lignocellulose, i.e. cellulose and hemicellulose are first depolymerized through pretreatment and saccharification and the obtained monomeric sugars are subsequently fermented to ethanol. Lignocellulose conversion to bioethanol can be carried out in various manners, such as by employing biochemical/microbial/enzymatic route or by adopting thermochemical/chemical route, however, following are some common considerations that need to be taken care of (Kang et al. 2014):

- Complete or near complete conversion of holocellulose components to respective monomeric sugars
- Improved co-fermentation in presence of pretreatment derived toxins
- Integration of unit operations for minimal waste generation and maximum energy utilization
- Lignin valorization to increase the cost-competitiveness of bioethanol production process

Despite various reports on sustainable production of cellulosic bioethanol via enzymatic route, a common argument against biofuels production is their high production costs. In this regard, many countries are providing governmental subsidies and tax exemptions to biofuels in order to achieve economic competitiveness against oil-derived transportation fuels. Moreover, the whole process could be made more cost-effective by generating high-value products from side streams in an integrated biorefinery manner, especially finding better alternatives of utilizing lignin for value-addition in comparison to its conventional application in heat generation (Balat et al. 2008).

The chapter aims to provide a concise overview of the basic concepts and newer developments as well as challenges and prospects of the state of the art related to the production of second generation biofuels. Various process steps in principle are discussed briefly and emphasis has been given on the advancement in each process step and to the challenges faced by the industries to make it commercially viable. Moreover, improvement of lignocellulose to bioethanol conversion process through genetic engineering approaches and development of biomass-based biorefinery has also been discussed.

2 Global Status of Second Generation Bioethanol Production

Advanced biofuels production over the world has been on rise since past few decades reaching more than 2×10^8 gallons annual production capacity and further developments and enhancement of production capacity in major biofuel producing nations is expected to nearly double current annual capabilities. Topmost nations on the list of global bioethanol producers are the United States, Brazil and China. The status of bioethanol production across the world is shown in Fig. 2. It is only recently that many advanced biofuel production plants, both demonstration as well as commercial scale, have been set up worldwide (US EPA 2016).

Different countries are using various substrates for bioethanol production depending on their regional availability, local climate and economic drivers. For example, in the US and Brazil, sugars derived from 1G resources such as maize and cane, respectively, are being used for ethanol production, whereas, China is using corn, wheat and sugarcane for production of bioethanol (Cardona and Sanchez 2007). The main drivers for biofuel development in India are secured energy supply by replacement or reduced usage of petroleum-based fuels. Indian biofuel policy targets to achieve 20% (volume) biofuels blending in fossil fuels by committing to establish various bioethanol and other advanced biofuel generation facilities in the whole country over a period of time and replacing the current sugar based substrates with lignocellulosic feedstock. A list of various first and second generation substrates currently used or proposed for bioethanol production is given in Table 1.

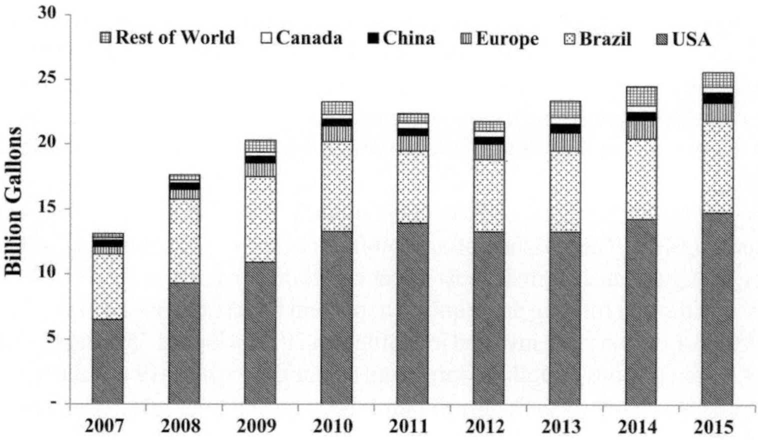

Fig. 2 Global trend for ethanol production (*Source* RFA 2015 www.ethanolrfa.org)

Table 1 List of various feedstock used for ethanol production in major biofuel producing countries

S. No.	Countries	Feedstock	
		First generation	Second generation
1.	US	Corn, Sugar beet	N.A.
2.	China	Corn, Wheat, Sugarcane	N.A.
3.	Germany	N.A.	Rye
4.	Brazil	Corn, Sugarcane	N.A.
5.	France	Sugar beet	N.A.
6.	Argentina	Soybean	N.A.
7.	Nigeria	Palm	Sorghum
8.	India	Wheat, Sugarcane	Sorghum
9.	Poland	N.A.	Rye
10.	Russia	Sugar beet	Rye
11.	Malaysia	N.A.	Palm waste
12.	Indonesia	Sugarcane molasses	N.A.
13.	Sudan	N.A.	Sorghum
14.	Columbia	Sugarcane	N.A.

Adapted and modified from (Araújo et al. 2017)
N.A. Not available

Since 2000, the global biofuels supply has increased by a factor of 8% and equalled 4% of the world's transport fuels in 2015. Global biofuels supply has improved enormously over past few years mainly due to adoption of biofuel policies by various countries with their own targets and mandates. The top two world-leading bioethanol producing countries have alone produced more than 1/3rd of the global bioethanol in 2015. USA has committed to increase its biofuel production capabilities to a level of approximately nine times of the current scenario and the European Union target to increase biofuel/bioenergy share by more than 10% by the year 2020 (Yacobbi 2012; (http://ec.europa.eu/energy/renewables/biofuels/biofuels_en.htm), accessed on 20th Dec 2017).

The advanced biofuels commercialization is more expensive than original expected biofuels. The absence of any biofuel policy worldwide is the major concern regarding the decline in the cost of per barrel oil prices from June 2014 to 2015. Necessary time and funding are required to prevent the decline in the biofuel market. Globally, $3.1 billion were invested in biofuels in 2015, which is 35% decline relative to 2014. Later, billions of dollars were spent on various projects of advanced biofuels worldwide, but many of such projects have been closed after sometime mainly due to commercialization issues (www.worldenergy.org).

3 Second Generation Bioethanol Process

Second generation bioethanol can be considered more environmentally friendly. Lignocellulosic biomass can either be by-products of agro-based industries and comprises sugarcane bagasse, rice straw, rice husks, wheat straw, cotton stalks, corn cob, coconut shells and municipal solid waste (MSW), forestry waste counting bark and wood chips. Lignocellulose is mainly made up of cellulose (polymer of α-D-glucose), hemicellulose (heteropolymer of C5 and C6 sugars) and lignin (heteropolymer of phenylpropanoid units). Numerous lignocellulosic biomasses can be successfully utilized for producing bioethanol. Some of them with their compositions are listed in Table 2. The process of lignocellulosics to ethanol broadly comprises of four sequential steps; Deconstruction of biomass (pretreatment), saccharification, conversion of sugar to ethanol (fermentation) and purification of the product (Fig. 3).

3.1 Pretreatment: Deconstruction of Lignocellulosic Biomass

Production of bioethanol from second generation biomass requires efficient depolymerization of structural carbohydrate polymers to be fermented to ethanol. However, lignocellulosic biomass has evolved complex structural and chemical mechanisms, which provide recalcitrance to its structural sugars from the microbial and enzymatic attack. Therefore, a deconstruction of biomass is required to change the biomass size and structure as well as chemical composition so that hydrolysis of the carbohydrate portion to monomeric sugars can be attained rapidly with higher yields. The main aim of pretreatment is as follows:

Table 2 Composition of various substrates used for bioethanol production

Substrate	% Composition (dry wt.)			Substrate	% Composition (dry wt.)		
	Hexosans	Pentosans	Lignin		Hexosans	Pentosans	Lignin
Bamboo	49–50	18–20	23	Oat straw	41	16	11
Banana waste	13.2	14.8	14	Olive tree waste	25.2	15.8	19.1
Barley hull	34	36	19.3	Paper	85–99	0–5	0–15
Barley pulp	69.9	18.3	10.9	Pepper stalks	35.7	26.2	18.3
Bean stalks	31.1	26.0	16.7	Pine	41	10	27
Bermuda grass	25	35.7	6.4	Poplar	40	14	20
Birch wood	40	33	21	Reed	49.40	31.50	8.74
Chilli stalks	37.5	28.3	17.3	Rice husk	36	15	19
Coffee pulp	33.7–36.9	44.2–47.5	15.6–19.1	Rice straw	32	24	13
Corn cobs	42	39	14	Rye straw	31	25	7
Corn stover	38	26	19	Salix	41.5	22–25	25
Cottonseed hair	80–95	5–20	0–5	Sawdust	55	14	21
Cotton stalks	41.7	27.3	18.7	Softwood stem	45–50	25–35	25–35
Douglas fir	35–48	20–22	15–21	Sorghum straw	33	18	15
Eucalyptus	45–51	11–18	29	Soybean stalks	34	25	20
Flax sheaves	35	24	22	Spruce	45	26	28
Grapevine stems	43.1	19.4	26.6	Sugarcane bagasse	33	30	29
Grasses	25–40	35–50	10–30	Sweet sorghum	23	14	11
Groundnut shells	38	36	16	Switch grass	37	29	19
Hemp	53.86	10.60	8.76	Waste paper	60–70	10–20	5–10
Jute fibres	45–53	18–21	21–26	Water hyacinth	18.4	49.2	–
Miscanthus	43	24	19	Wheat straw	30	24	18
Municipal solids	8–15	NA	24–29	Willow	55.9	14	19

Sources Monsalve et al. 2006; Karp and Shield 2008; Kim et al. 2008; Alves et al. 2010; Singh et al. 2011; Garcia 2014; Ayeni et al. 2015; Raud et al. 2016; Bilal et al. 2017; Espinosa et al. 2017

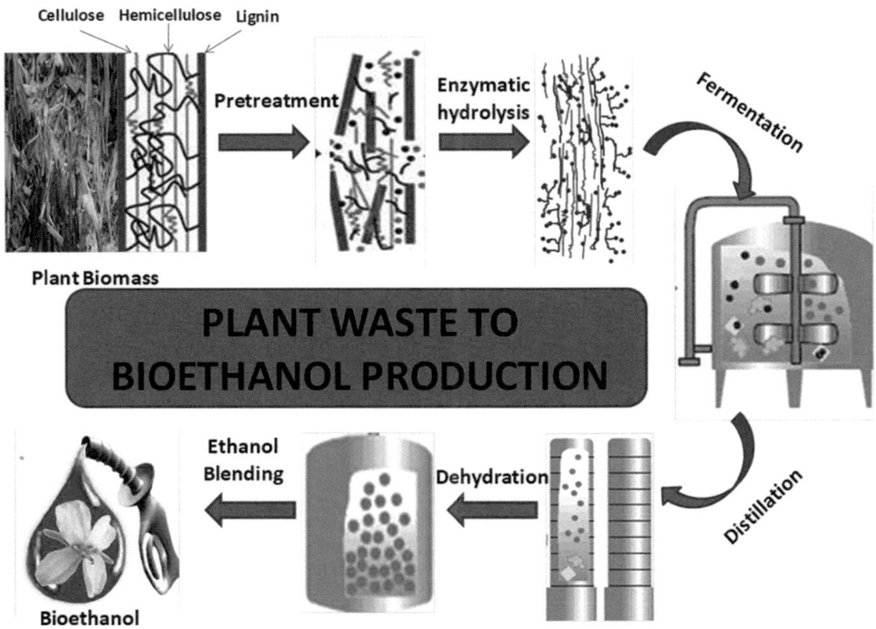

Fig. 3 Schematic illustration of second generation bioethanol production process

(1) To improve sugar yields during enzymatic hydrolysis by reduction of crystallinity of cellulose and enhanced porosity of the biomass;
(2) To minimize the emergence of fermentation inhibitors during deconstruction;
(3) To retrieve lignin from hydrolysate for converting it into valuable by-products and
(4) To make the process economic by making the operation easier (Aditiya et al. 2016).

Broadly pretreatment strategies are categorized into physical, physico-chemical, chemical and biological. With every different feedstock used for bioethanol production, the selection of pretreatment method varies due to distinct chemical composition and physical structure of feedstock. Factors like cellulose crystallinity, lignin content, cell wall porosity, hemicellulose side chain branching and crosslinking are critical in choosing the pretreatment method. Most chemical pretreatment modifies cellulose ultrastructure through certain physico-chemical modification, though it is possible to fractionate cellulose, hemicellulose and lignin by using pretreatment with some catalysts.

Pretreatment using acids or bases promote subsequent enzymatic hydrolysis by exposing cellulose and removing hemicellulose consequently enhancing the yield of glucose. The frequently used acid and base are H_2SO_4 and NaOH, respectively. Another additive, cellulose solvents have been used to liquefy cellulose in various cellulosic substrates which ultimately results in 90% conversion of cellulose to glucose and substantiated raised enzymatic hydrolysis due to the deconstruction of biomass before the action of enzyme. Organosolvants like Lewis acids, $FeCl_3$ and $(Al)_2SO_4$, and alkaline-peroxide (H_2O_2) are known solvents to disintegrate lignocellulosic structure and facilitates hydrolysis (Coughlan 1992). Concentrated acids such as sulphuric acids (H_2SO_4) and hydrochloric acid (HCl), alkali solvents like NH_3 and hydrazine, aprotic solvents (DMSO), and some complexes of metal and wet oxidation enhance the porosity of biomass by interrupting the association of lignin with cellulose and also dissolving hemicellulose. Although the abovementioned methods are effective, the cost of these chemicals is high when compared with the value of the glucose and hence make their use impractical (Sun and Cheng 2002).

Besides, high temperature/pressure-based pretreatments and biological pretreatment with selected lignin degrading white rot fungi have been used successfully. Contrary to chemical based methods, input of energy in biological pretreatment is lesser as the reaction conditions are milder. White rot fungi can effectively degrade lignin by secreting hydrolases with lignin peroxidases which in the presence of H_2O_2 cleaves the backbone of lignin. A list of common pretreatment strategies used and their advantages and disadvantages are shown in Table 3.

This is interesting to note that while performing chemical-based pretreatments, generation of various fermentation inhibitors (furfural, hydroxymethyl furfural, phenolics, acetic acid, etc.) takes place. Therefore, prior to fermentation, removal of these inhibitors seems necessary. Several detoxification strategies such as liming, activated charcoal adsorption, ion-exchange resin treatment and enzymatic detoxification have been used to remove these fermentation inhibitors. An alternative and more sustainable way to tackle the problem of inhibitors is to use inhibitor resistant or tolerant enzymes and microbial strains.

3.2 Enzymatic Hydrolysis: Depolymerization of Structural Polymers

The hydrolysis of pretreated biomass is the most crucial step in the bioethanol production process. Although hydrolysis of biomass can be accomplished by using acid or enzymes, saccharification using enzymes is preferred due to milder processing conditions and environment-friendly nature. Nevertheless, depolymerization of biomass via enzymatic hydrolysis is a multi-enzymatic process with high complexities.

In nature, lignocellulosic biomass can be depolymerized by a number of hydrolytic enzymes that are produced by diverse fungi and bacteria. Cellulases are the representative class of enzymes involved in depolymerizing lignocellulosic substrate by synergistic action of all the three enzymes present in the complex. Cellulase complex consists of exoglucanases (cellobiohydrolases, CBH), endoglucanases (EG) and β-glucosidases (cellobiase, BG) (Behera and Ray 2016). EG acts upon cellulose chains and hence creates two types of reactive ends for CBHs. CBH I acts on reducing ends and CBH II on non-reducing ends of cellulose fragments thereby, catalysing step-

Table 3 Various pretreatment strategies with their specifications (Aditiya et al. 2016)

Pretreatment	Action	Advantages	Disadvantages
Dilute acid	Hydrolyses hemicelluloses, Concentrates cellulose enzymatic treatment, Alters lignin structure	Hemicellulosic removal	Low removal of lignin, low enzymatic hydrolysis (30–40%), Inhibitor generation
Dilute alkali	Eliminates lignin and hemicelluloses, Surface area exposed for enzyme access	High digestibility, high lignin removal	Hemicellulosic sugar loss, Low enzymatic hydrolysis (50–60%), Inhibitor generation
Ammonia fibre expansion (AFEX)	Surface area for access to enzyme upsurge after treatment Take out hemicellulose and lignin	Small amount of inhibitors formation	Not proficient for biomass with high level of lignin, High price of ammonia
Ionic-liquids	Increases proportion of amorphous cellulose, Lignin is separated	High dissolution, Environmentally safer	Scale-up is still a challenging
Alkaline peroxide	Removes lignin and solubilize most of the hemicellulose	Cellulose isolation	Loss of hemicellulosic sugars, Loss of lignin
Acid-chlorite	Reduces lignin content	Isolation of hemicellulose and cellulose	Loss of lignin, Costly method of pretreatment
Ammonia	Opens up cell wall and exposes celluloses and hemicelluloses	Lignin removal (partial)	Hemicellulosic sugar loss, Low enzymatic hydrolysis (50–60%), Inhibitor generation
Steam-explosion	Porosity of biomass increases, Hemi-cellulose solubilization	Deconstruction of structural polymer, Recovery of lignin, Lower loss of hemicellulose, Less amount of inhibitors generated, Higher yield of hemicellulose and, economic process	Generation of inhibitors, Partial degradation hemicellulosic components, Disrupted lignin-carbohydrate matrix is lacking
Biological	Ligninolytic and hemicellulolytic action	Partial deconstruction of lignocellulosics, Low energy consumption	Longer fermentation time
Lignin downregulation	Development of transgenic plants with downregulated lignin	Lower lignin content plants	Susceptible to disease

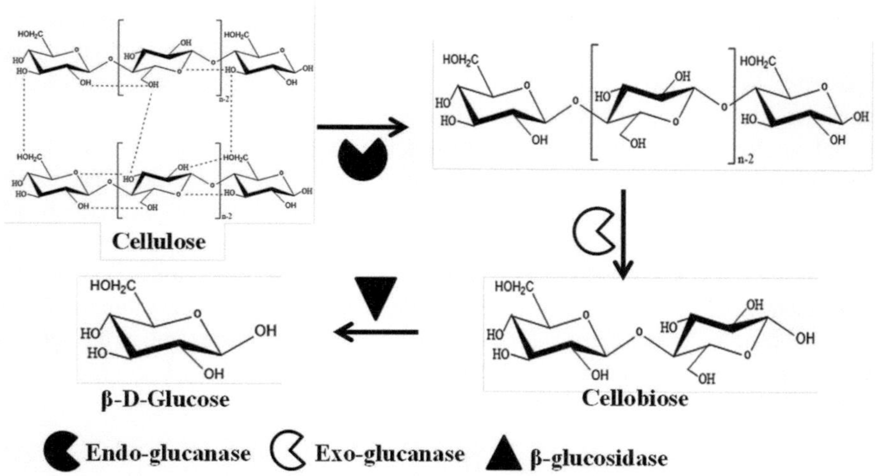

Fig. 4 Schematic diagram showing mechanism of enzymatic hydrolysis

wise degradation of cellulose to cellobiose. BG utilizes cellobiose and converts it into glucose (Kuhad et al. 2011b). CBH gets inhibited by cellobiose, therefore; BG plays a key role in reducing end-product inhibition and depolymerizing the cellulose completely. Modular structure with concluding catalytic and carbohydrate binding molecules (CBM) is the common feature of most of the cellulases. The carbohydrate binding molecules facilitate hydrolysis of biomass by fetching the catalytic domain in contiguity to the insoluble cellulose. Thus, the rate of enzymatic hydrolysis of the biomass is subjective to the substrate properties and catalytic performance both. The scheme of mechanistic action of cellulases over cellulose is shown in Fig. 4.

Besides cellulases, several other auxiliary enzymes such as xylanases, mannases, feruloyl esterases, etc. also assist the enzymatic depolymerization of lignocelluloses. Recently, novel enzymes (non-hydrolytic) named lytic polysaccharide monooxygenases (LPMOs) have been reported to be capable in dropping cellulase dosages and finally the overall cost of the process (Vaaje-Kolstad et al. 2010; Horn et al. 2012). Although the mechanism is not clear yet, these LPMOs are believed to oxidize the highly recalcitrant crystalline regions of cellulose and create more reducing/non-reducing ends for cellulase components to attack (Horn et al. 2012). This may be due to the fact that LPMOs require an electron donor, e.g. oxygen, for their effective action (Hu et al. 2015).

Although saccharification using enzymes has more scope for improvements than those using chemicals, the high cost of cellulases is still a technical barrier (Hong et al. 2013; Culbertson et al. 2013). Fall in the cost of cellulase could be obtained by (a) intensive efforts which enquire more than a few aspects of enzymes with improved hydrolytic properties such as binding affinity, thermostability, etc. (b) by improvement of technologies for which are proficient for hydrolysis including of superior cocktails of enzyme and conditions for hydrolysis. In addition to enzyme characteristics, substrate features such as the degree of polymerization, cellulose crystallinity and the existence of lignin and hemicellulose also affect the enzymatic hydrolysis. Therefore, to improve the overall process, upgrading in cellulase performance and enhancing the substrate-enzyme interaction are prerequisite.

Industrially, among all probable strategies, the optimization of the characteristics of cellulases like thermostability and end-product inhibition is crucial for large-scale application. Also, optimizing production medium by altering its components is an approach to enhance the enzymatic hydrolysis. Development of multi-enzyme cocktail secreted by various strains of fungi is also a good choice for improving the performance of cellulase as a complete system. Several studies have reported that synergistic action of cellulase is linked with the ratio of every enzyme in the system (Berlin et al. 2007; Hemansi et al. 2018).

3.3 Fermentation

As compared to simpler fermentation process of sugars derived from food-based feedstock, crop-waste based feedstock to ethanol conversion process is very tedious and involves many critical steps. Pentose-rich sugar syrup and hexose rich sugars coming from hydrolysis of hemicellulose and cellulose, respectively, are the major substrates after initial hydrolysis that can be further fermented to produce ethanol. There are many desirable characteristics of an ideal fermenting microorganism, such as high conversion efficiency both with respect to substrate utilized and time, robustness against inhibitory compounds and ability to withstand high ethanol concentrations.

Several laboratories have established the process of utilizing pentose sugars as well as hexose sugars by various yeasts, fungi and bacteria for the production of fermentation products including alcohols (Tables 4 and 5. Among these, the most common and efficient glucose fermenting microbes are brewer's yeast and *Zymomonas mobilis* (Hahn-Hägerdal et al. 2006), while for pentose fermentation are *Pichia stipitis* and *Candida shehatae*.

The process of ethanol production not always requires aerobic conditions. It is required only for the production of biomass (Agbogbo and Wenger 2007).

Further to enhance the ethanol production from pentose sugars, different detoxification strategies have been used by various researchers (Chandel et al. 2007). The elimination of inhibitors from fermentation broth considerably improved the yield and productivity of ethanol as compared to un-detoxified hydrolysate. Moreover, utilization of all the sugars including hexoses (C6; glucose, galactose, and mannose) and pentoses (C5 sugars; xylose and arabinose) in a single reactor can be another option to reduce the cost of producing cellulosic bioethanol.

Scientists around the world have employed different fermentation strategies for cost-effective processes for ethanol production from lignocellulosic biomass in a single reactor. These processes include separate hydrolysis and fermentation (SHF), simultaneous saccharification and fermentation (SSF), simultaneous saccharification and co-fermentation (SSCF), consolidated bioprocessing (CBP), and simultaneous

Table 4 Various pentose fermenting microorganisms

Bacteria	References	Fungi and yeasts	References
Klebsiella oxytocea	Ingram et al. (1999)	*Neurospora crassa*	Deshpande et al. (1986)
Lactobacillus pentosus	Chaillou et al. (1999)	*Pachysolen tannophilus*	Schneider et al. (1981)
Lactobacillus casei	Roukas and Kotzekidou (1997)	*Paecilomyces* sp NF1	Mountfort and Rhodes (1991)
Lactobacillus pentoaceticus	Chaillou et al. (1999)	*Pichia stipitis*	Gupta et al. (2009)
Lactobacillus plantanum	Sreenath et al. (1999)	*Rhizopus orizae*	Millati et al. (2005)
Lactobacillus xylosus	Sreenath et al. (1999)		

Table 5 Various hexose fermenting microorganisms

Hexose fermenting microorganisms			
Organisms	References	Organisms	References
Fusaruium sporium	Mamma et al. (1995)	*Rhizomucor pusillis*	Millati et al. (2005)
Kloeckera apiculata	Aguilera et al. (2006)	*Saccharomyces cerevisiae*	Kuhad et al. (2010)
Kluyeromyces marxianus	Ballesteros et al. (2004)	*S. bayarus*	Belloch et al. (2008)
Mucor indicus	Abdenifar et al. (2009)	*S. paradoxus*	Belloch et al. (2008)
Pachysolen tannophilus	Abbi et al. (1996)	*S. pastorianus*	Belloch et al. (2008)
Pichia stipitis	Gupta et al. (2009)	*Schizosaccharomyces pombe*	Hu et al. (2005)
Pichia membranifaciens	Aguilera et al. (2006)	*Terulospora delbruecki*	Aguilera et al. (2006)
Rhizopus oryzae	Abdenifar et al. (2009)	*Zymomonas mobilis*	

saccharification, filtration and fermentation (SSFF). All the processes have been shown in Fig. 5.

Above mentioned methods (SSF, SSCF, CBP) are preferred over separate enzymatic deconstruction and fermentation (SHF) strategy. Despite it, in current scenario, SHF is the mostly used method for bioethanol production. During the first step of SHF, cocktail of lignocellulolytic enzymes is produced so that lignocellulosic biomass can be converted into a syrup of monomeric sugars (hexoses/pentoses). This solution is further used to produce bioethanol with the help of pentose/hexose fermenting microbes in a separate step. For the first step, i.e. hydrolysis, optimum temperature ranges from 45 to 50 °C, whereas for fermentation, the optimal range is near 30 °C, so both steps are performed sequentially. In SSF, enzymatic hydrolysis of pretreated lignocellulosic biomass to release monomeric sugars for subsequent

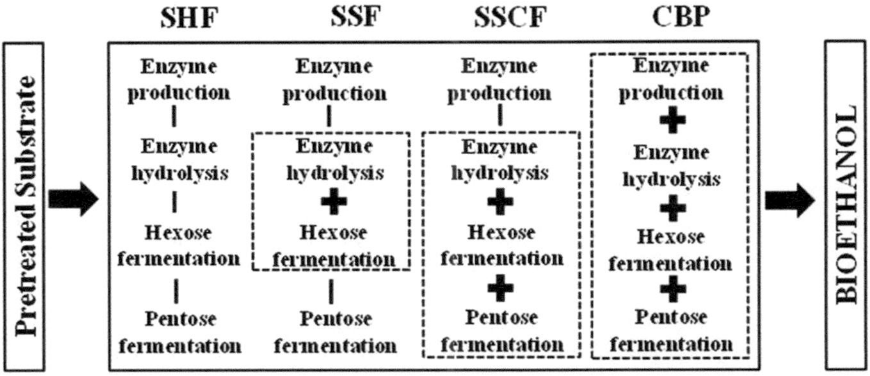

Fig. 5 Overview of various fermentation strategies

microbial conversion to ethanol is performed in the same vessel. Hallmark of this type of process is the compromise between optimum temperatures of both hydrolysis and fermentation (Choudhary et al. 2016). SSF is important over SHF as it delimits repression of cellulases (by glucose) via feedback inhibition, so improves the efficiency of saccharification as well as ethanol yield.

Further improvements in the ethanol titres and yields can be achieved if saccharification and simultaneous conversion of both five- and six-carbon sugars can be carried out (SSCF method) depending upon the fermentation capacity of the microorganisms. During the process, cellulases feedback inhibition also gets inhibited in a similar way to that of SSF, enhancing the efficiency of co-fermentation. Consolidated bioprocessing (CBP) is a relatively newer process configuration in which various biomass conversion steps such as synthesis of lignocellulolytic enzymes, feedstock deconstruction and final conversion to ethanol are performed in an integrated manner by a single microorganism. It is a comparatively promising, long-lasting and cost-effective approach for ethanol production, because of lesser requirements than other process configurations. However, current research shows that CBP-based configuration is still in its infancy and there is a lot of scope for the development of better and robust CBP organisms through molecular biology and recombinant DNA-based approaches.

SSFF is another integrated process where saccharification and fermentation chambers are separated by a membrane filtration chamber. Most of the genetically engineered or natural yeasts do not efficiently convert hexoses as compared to pentose conversion and thus, fermentation of pentose begins after that of hexose sugars. SSFF is more efficient in comparison to separate or simultaneous fermentation approaches as it provides conditions for hydrolytic enzymes and the fermenting microbes that can be maintained separately. In brief, hydrolytic enzymes carry out hydrolysis in a separate chamber and are filtered and recycled back using a tangential flow membrane filtration system. The filtrate rich in sugar is further put back into the compartment where final fermentation can take place chamber and hence, both the chambers are maintained at similar working capacity. Furthermore, applications of flocculating yeasts can help in cell harvesting and recycling of the settled microbial biomass in the fermenter (Ghose and Bandyopadhyay 1980).

4 Genetic Engineering Approach for Bioethanol Process Improvement

Yeasts belonging to genera like *Saccharomyces, Candida, Kluyveromyces, Pachysolen, Pichia, Brettanomyces* and *Schizosaccharomycesetc* are used for bioethanol production. Out of these, *Saccharomyces cerevisiae* is commonly employed in bioethanol production due to higher productivity, high ethanol tolerance and the ability of fermenting hexoses rapidly. However, it cannot utilize pentose sugar (mainly xylose) due to the absence of key enzymatic machinery required for pentose sugar metabolism. There are two pathways present naturally among fungi and bacteria. The pathway present in fungi utilizes xylulo-reductase (XR) and xylose dehydrogenase (XDH) enzymes for the conversion of D-xylose into its isomer D-xylulose while another pathway present in bacterial utilizes xylulo-isomerase (XI) that converts the same in single step. Xylulose then enters the pentose phosphate pathway in the form of xylulose-5-phosphate by the activity of the enzyme xylulo-kinase (XKS) common to every pathway for sugar metabolism (Fig. 6).

Though *S. cerevisiae* harbours XKS gene but does not have XR/XDH (XYL1/XYL2) or XI (Xyl a) gene. Besides, various transporters are also needed for the entry of pentose sugar. Various combinations of these key genes in vector based transformation and genomic integration have been widely attempted for higher production of ethanol (Table 6) but these combinatorial approaches pose two major limitations: (a) Xylitol Accumulation: the main problem of XR–XDH pathway is incomplete recycling of redox co-substrates ($NADPH/NAD^+$) during catalysis of NADPH dependent XR and the NAD^+ preferring XDH which forms xylitol (a valuable by-product) and hence it lowers overall yield of ethanol from xylose. (b) Lower catalytic efficiency of XI. One practical solution can be the replacement of XI by XR-XDH pathway to overcome cofactor preference, but its catalytic efficiency is much lower and slower.

These limitations have been addressed by (i) XR mutation for preference of cofactor via genetic engineering for higher ethanol yield, and (ii) improvement in its genetic makeup or codon optimization. Moreover, it was also observed that overexpression of XKS1 and TAL1 (transaldolase), TKL1 (transketolase), RPE1 (ribulose5-phosphate epimerase) and RKI1 (ribose 5-phosphate keto-isomerase) (Genes of Non-oxidative pathway of *S. cerevisiae*) may lead to enhanced production of ethanol and reduction in xylitol production. In addition, enhancement in ethanol production could be accomplished by decreasing glycerol, the main side-product during glucose fermentation. In *S. cerevisiae*, GPD1 and GPD2, two ample NAD-dependent glycerol-3-phosphate dehydrogenases, are main enzymes in the synthesis of glycerol with NADH produc-

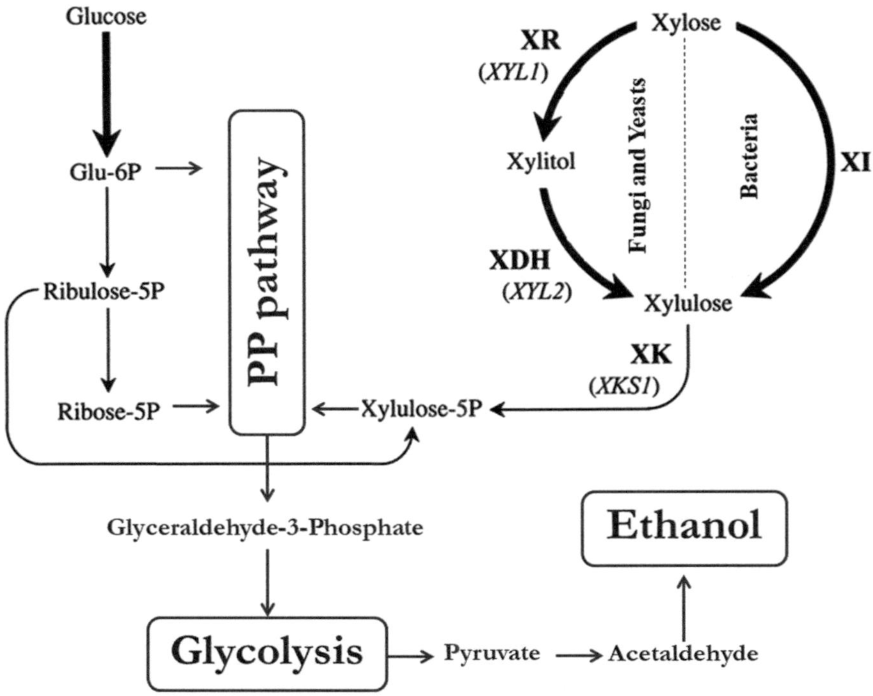

Fig. 6 Schematic diagram for xylose fermentation pathway for ethanol production

tion. This problem can be addressed by deletion of glycerol metabolism genes and overexpression of genes of glutamate pathway (Glutamate synthase, GLN1 or GLT1), which can increase ethanol production and reduce glycerol production significantly.

5 Energy and Mass Balance for Cellulosic Ethanol Production

The application of energy and mass balance regularities appear to be useful for the estimation of the efficiency of bioethanol production. The use of lignocellulosic biomass feedstock will markedly decrease energy input/output ratio. During the process of bioethanol production, the feedstock runs through a series of process steps and to make the process sustainable and economic, maximum output of energy should be attained. For this, energy inputs and outputs at every step of typical ethanol production process are analysed. Moreover, a detailed analysis of mass balance should be prepared. A schematic diagram of typical bioethanol production process comprising acid pretreatment, detoxification, delignification, pentose fermentation and hexose fermentation under SHF fermentation strategy is shown in Fig. 7.

Table 6 List of few pentose fermenting recombinant *S. cerevisiae* and their ethanol production potential

Strain	Sugar used (g/L)	Ethanol yield (g/g)	Ethanol productivity (g/L/h)	References
S. cerevisiae TMB 3001	G:X 5:15	25	0.15	Eliasson et al. (2000)
S. cerevisiae TMB 3001	10 X	48	NA	Sonderegger and Sauer (2003)
S. cerevisiae F12	G:X 50:50	52	NA	Sonderegger et al. (2004)
S. cerevisiae TJ1	50 X	10.6	0.02	Tantirungkij et al. (1993)
S. cerevisiae TMB 3001	10 X	88	0.061	Träff-Bjerre et al. (2004)
S. cerevisiae H 2673	50 X	46	NA	Verho et al. (2003)
S. cerevisiae ZU-10	80 X	75.6	0.50	Zhao and Xia (2009)

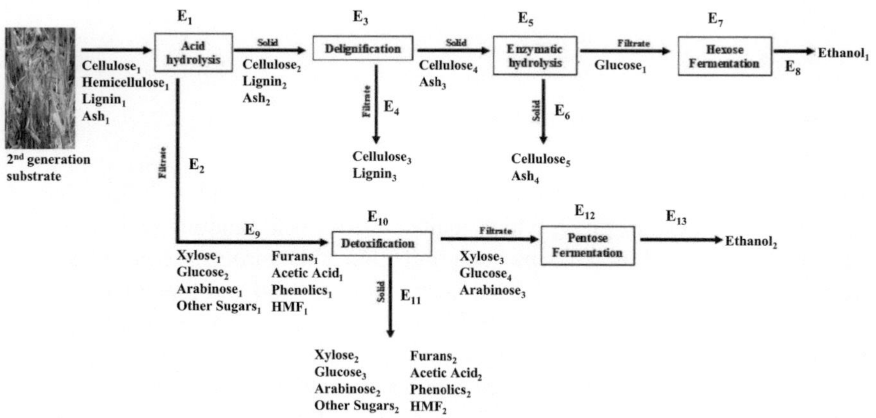

Fig. 7 Schematic overview of various steps for energy and carbon evaluation in a process for production of bioethanol from second generation feedstock using SHF strategy

6 Life Cycle Analysis or Assessment of Cellulosic Ethanol Production Processes

Besides manufacturing expenses and method, which determine the overall economic sustainability, various environmental and social criteria must also be considered for designing the biofuel production process (IEA technology 2011). Few regulatory as well as volunteer bodies (GBP 2011; ISO 2009; RSB 2012) have been instrumental in formulating set of standards and benchmarks for sustainable biofuel manufacturing. Many reports in the literature are available as far as the comparison of the socio-econo-environmental sustainability aspects of second generation bioethanol production processes is concerned.

The LCA is an assessment of contributions and productions to determine the effect of products formed on environment throughout the life cycle. LCA is created to compare the impacts of a product, process and/or service to generate environmental awareness in customers, governments and companies (ISO 2006). LCA could also be functional to evaluate improvement in product, its designing and comparison. It considers four phases: (a) Defining limits and objectives of system, (b) to access the inventory of life cycle, (c) quantification of life cycle impacts, and (d) results interpretation (Morales et al. 2015).

Environmental effects target to enumerate the effect of global warming, ozone depletion, photochemical oxidation and others (Roy et al. 2012). A number of software such as SimaPro, LCAmanager, Umberto, etc. have been designed to help assessing the LCA, which also involves database from various economic sectors, which may differ in their quality. These software quantify the effects of emissions on different objectives and are in favour for the different effects like depleting ozone layer, eutrophication, global warming, etc. LCA analysis of few commonly used substrates in different countries is listed in Table 7.

7 Techno-economic Evaluation

Stone & Webster Engineering Corporation (1987) evaluated the feasibility of wood-based cellulosic ethanol plant, which includes feedstock handling, acid catalysed steam explosion pre-treatment, enzyme production and hydrolysis, concentration of glucose, fermentation, distillation and anaerobic digestion and the ethanol selling price was estimated to be $0.93/l or $3.5/gal. Similarly, another report released by Chem Systems, Inc. (1987) which consisted of separate saccharification and ethanol fermentation of hardwood, enzyme production, CO_2 recovery and furans production, estimated an ethanol selling price of 0.54/l or $2.06/gal. Later on, NREL reported the lignocellulose conversion to ethanol following acid hydrolysis at a cost of ~$0.05/l or $ 0.20/gal ethanol. They also reported that though enzymatic hydrolysis has great potential for improvement, the saccharifying enzymes are very expensive (~US$0.08–0.13/l ethanol or 0.3–0.5/gal ethanol) (Aden et al. 2002). In the past decades, maximum efforts were focussed to reduce the enzyme production cost. Aden et al. (2002) estimated that if the enzyme cost comes less than 2.67 cents/l or 10 cents/gal ethanol, the cost of ethanol production could drop as low as $0.28/l or $1.07/gal and in another report NREL has aimed to achieve this goal by 2012 (Aden 2008). Concerning the R&D in cellulosic ethanol, a multi-year program was planned, which has to be updated every 2 years, including 2005 (US DOE, 2005), 2007 (US DOE, 2007) and 2009 (US DOE, 2009). The detailed updates of the technology model are provided by Aden and Foust (2009). In the European Commission, seven EU institutes evaluated the biofuels potential and costs (Hamelinck et al. 2005;

Table 7 LCA of 1G and 2G bioethanol processes

S. No.	Substrate	Type of feedstock	Growth time (Months)	Growth temperature	Water requirement	g CO_2/MJ	Energy balance	References
1.	Corn	1st	4–5	18–20 °C	50–80 cm	90	1.25	Thompson (2012)
2.	Sugar beet	1st	4–6	20–25 °C	55–75 cm	30–106	1.4–2.1	Araújo et al. (2017)
3.	Sugarcane	1st	15–16	32–38 °C	150–250 cm	5.06–59.3	4.4	Araújo et al. (2017)
4.	Wheat	1st	3–4	15–20 °C	45–65 cm	56–77	1–1.2	Araújo et al. (2017)
5.	Lignocellulose	2nd	3–4	ND	Very little	4–40	1.23–2.2	Araújo et al. (2017)
6.	Agriculture waste (Sorghum, Rye straw)	2nd	3.5–5	20–35 °C	45–65 cm	23–72	0.78–1.79	Araújo et al. (2017)

Table 8 Major technological bottlenecks in bioethanol development process

Pretreatment	Enzymatic hydrolysis	Fermentation
• Single or universal pretreatment • Lignin recovery • No inhibitor generation • Efficient Conditioning Strategy • Recovery or reuse of input energy • Recovery or reuse of used water Fully integrated process	• Availability of low-cost enzyme • Development of substrate-specific enzyme formulation • Specially designed reactor for high substrate consistency • Capability of converting unreacted xylan/xylo-oligomers • Operation in whole slurry mode (inhibition tolerance)	• Inhibitor tolerant microbes • The approach of SSF or CBP should be used • Bioprospecting for efficient pentose fermenting strain • Efficient conversion of hemicellulose sugars to other value-added product such as xylitol • Genetically modified strain for mixed sugar fermentation

Gnansounou and Dauriat 2010). The economic evaluation took into account the manufacturing cost of \$0.90/l or €0.62/l in 2010, \$0.85/l or €0.59/l in 2020 and \$0.72/l or €0.50/l in 2030. In another case study, Sassner et al. (2008) compared the economic performances for the conversion of different lignocellulosics (Spruce, corn-stover and salix) to ethanol, which required estimation of annual production cost including annualized capital cost and annual operation costs. According to them, the annual production costs (US\$) vary significantly, i.e. \$0.66–0.69/l ethanol (spruce), 0.67–0.86 (corn stover) and 0.72–0.87 (salix). Reports on LCA of cellulosic bioethanol from Indian researchers are very few in comparison to other countries.

8 Future Prospects

Development of cellulosic ethanol as a biofuel is very much needed at present, as it will have the potential to make countries self-sufficient in the energy sector and make the environment more safer and greener. Globally, the focus has already shifted from food-based resources towards non-food crop wastes (Saha et al. 2005; Himmel et al. 2007; Kuhad et al. 2011a; Saini et al. 2015). However, to reduce the final production costs, major cost-contributing steps have to be optimized from a technical as well as economical point of view (Table 8).

Priority should be on development of highly efficient and cheaper cellulolytic enzymes that can be produced economically and can act very fast even at a minimal dose. Additionally, an environmentally greener as well as cheaper and highly efficient pretreatment technology has to be used that will further reduce the efforts and costs in subsequent steps. As far as improvement of fermentation technology is concerned, there is still a very large scope for development of very robust and efficient pentose fermenting microorganisms. Priority should be developments in research and technological advancements in co-fermentation of hexose and pentose sugars simultaneously at a greater ease. It will definitely require more robust applications of molecular biology and metabolic engineering approaches (Galazka et al. 2010) as well as adjustments of metabolic flux (Matsushika et al. 2008). Another robust technology could be the development of consolidated bioprocessing microorganisms that have better catalytic abilities (Zhang et al. 2009). And finally, the successful transition of the lab or demonstration scale technologies to a large industrial scale will finally help in establishing commercial level cellulosic ethanol plants based upon currently available processes. In short, concerted efforts by experts from various science and technological disciplines will be required to tackle the hurdles that the current cellulosic ethanol industry is facing.

9 Conclusion

The potential to use lignocellulosic biomass from various sectors to produce second generation bioethanol underscores the need of technological advancement in each and every process step. The impediments of lower sugar recovery, hemicellulose fermentation, enzyme recycling, etc. need extensive inputs to be taken care of. The technological interventions for better biomass deconstruction strategies in conjunction with better process integration and optimization are required. One of the better strategies may be development and application of most efficient organisms in association with smart integration of various processes in an integrated biorefinery approach, where a multitude of products can be obtained in addition to bioethanol only and this may also include applications of consolidated bioprocessing microorganisms.

Acknowledgements The authors highly acknowledge financial and infrastructural support from Central University of Haryana, Mahendergarh, Haryana, India. The authors would also like to thank Department of Science and Technology (SERB-DST), Government of India for providing the financial support (SERB/LS/2016/00929).

References

Abbi M, Kuhad RC, Singh A (1996) Bioconversion of pentose sugars to ethanol by free and immobilized cells of Candida shehatae (NCL-3501): fermentation behaviour. Process Biochem 31:555–560

Abdenifar S, Karimi K, Khanahmadi M (2009) Ethanol production by *Mucor indicus and Rhizopus oryzae* from rice straw by separate hydrolysis and fermentation. Biomass Bioenergy 33:828–833. https://doi.org/10.1016/j.biombioe.2009.01.003

Aden A (2008) Biochemical production of ethanol from corn stover: 2007 state of technology model (No. NREL/TP-510-43205). National Renewable Energy Laboratory (NREL), Golden, CO (United States)

Aden A, Foust T (2009) Technoeconomic analysis of the dilute sulfuric acid and enzymatic hydrolysis process for the conversion of corn stover to ethanol. https://link.springer.com/article/10.1007/s10570-009-9327-8

Aden A, Ruth M, Ibsen K, et al (2002) Lignocellulosic biomass to ethanol process design and economics utilizing co-current dilute acid prehydrolysis and enzymatic hydrolysis for corn stover. National Renewable Energy Lab, Golden

Aditiya HB, Mahlia TMI, Chong WT et al (2016) Second generation bioethanol production: a critical review. Renew Sustain Energy Rev 66:631–653. https://doi.org/10.1016/j.rser.2016.07.015

Agbogbo FK, Wenger KS (2007) Production of ethanol from corn stover hemicellulose hydrolyzate using *Pichia stipitis*. J Ind Microbiol Biotechnol 34:723–727. https://doi.org/10.1007/s10295-007-0247-z

Agrawal R, Singh NR, Ribeiro FH, Delgass WN (2007) Sustainable fuel for the transportation sector. Proc Natl Acad Sci 104:4828–4833. https://doi.org/10.1073/pnas.0609921104

Aguilera F, Peinado RA, Millán C et al (2006) Relationship between ethanol tolerance, H^+-ATPase activity and the lipid composition of the plasma membrane in different wine yeast strains. Int J Food Microbiol 110:34–42. https://doi.org/10.1016/j.ijfoodmicro.2006.02.00

Alves EF, Bose SK, Francis RC et al (2010) Carbohydrate composition of eucalyptus, bagasse and bamboo by a combination of methods. Carbohydr Polym 82:1097–1101. https://doi.org/10.1016/j.carbpol.2010.06.038

Araújo K, Mahajan D, Kerr R, da Silva M (2017) Global biofuels at the crossroads: an overview of technical, policy, and investment complexities in the sustainability of biofuel development. Agriculture 7:32. https://doi.org/10.3390/agriculture7040032

Ayeni AO, Adeeyo O, Oresegun O, Oladimeji T (2015) Compositional analysis of lignocellulosic materials: evaluation of an economically viable method suitable for woody and non-woody biomass

Balat M, Balat H, Öz C (2008) Progress in bioethanol processing. Prog Energy Combust Sci 34:551–573. https://doi.org/10.1016/j.pecs.2007.11.001

Ballesteros M, Oliva JM, Negro MJ et al (2004) Ethanol from lignocellulosic materials by a simultaneous saccharification and fermentation process (SFS) with *Kluyveromyces marxianus* CECT 10875. Process Biochem 39:1843–1848. https://doi.org/10.1016/j.procbio.2003.09.011

Behera SS, Ray RC (2016) Solid state fermentation for production of microbial cellulases: Recent advances and improvement strategies. Int J Biol Macromol 86:656–669. https://doi.org/10.1016/j.ijbiomac.2015.10.090

Belloch C, Orlic S, Barrio E, Querol A (2008) Fermentative stress adaptation of hybrids within the *Saccharomyces* sensu stricto complex. Int J Food Microbiol 122:188–195. https://doi.org/10.1016/j.ijfoodmicro.2007.11.083

Berlin A, Maximenko V, Gilkes N, Saddler J (2007) Optimization of enzyme complexes for lignocellulose hydrolysis. Biotechnol Bioeng 97:287–296. https://doi.org/10.1002/bit.21238

Bilal M, Asgher M, Iqbal HMN et al (2017) Biotransformation of lignocellulosic materials into value-added products—a review. Int J Biol Macromol Complete 447–458. https://doi.org/10.1016/j.ijbiomac.2017.01.133

Cardona CA, Sanchez OJ (2007) Fuel ethanol production: Process design trends and integration opportunities. Bioresource Technol 98:2415–2457. https://doi.org/10.1016/j.biortech.2007.01.002

Chaillou S, Pouwels PH, Postma PW (1999) Transport of d-Xylose in *Lactobacillus pentosus*, *Lactobacillus casei*, and *Lactobacillus plantarum*: evidence for a mechanism of Facilitated diffusion via the Phosphoenolpyruvate: Mannose Phosphotransferase System. J Bacteriol 181:4768–4773

Chandel AK, Kapoor RK, Singh A, Kuhad RC (2007) Detoxification of sugarcane bagasse hydrolysate improves ethanol production by *Candida shehatae* NCIM 3501. Bioresour Technol 98:1947–1950. https://doi.org/10.1016/j.biortech.2006.07.047

Charles MB, Ryan R, Ryan N, Oloruntoba R (2007) Public policy and biofuels: the way forward? Energy Policy 35:5737–5746. https://doi.org/10.1016/j.enpol.2007.06.008

Choudhary J, Singh S, Nain L (2016) Thermotolerant fermenting yeasts for simultaneous saccharification fermentation of lignocellulosic biomass. Electron J Biotechnol 21:82–92. https://doi.org/10.1016/j.ejbt.2016.02.007

Claassen PAM, Lier JB van, Contreras AML, et al (1999) Utilisation of biomass for the supply of energy carriers. Appl Microbiol Biotechnol 52:741–755. https://doi.org/10.1007/s002530051586

Coughlan MP (1992) Enzymic hydrolysis of cellulose: an overview. Bioresour Technol 39:107–115. https://doi.org/10.1016/0960-8524(92)90128-K

Culbertson A, Jin M, Sousa L da C, et al (2013) In-house cellulase production from AFEX^{TM} pretreated corn stover using *Trichoderma reesei* RUT C-30. RSC Adv 3:25960–25969. https://doi.org/10.1039/c3ra44847a

Deshpande V, Keskar S, Mishra C, Rao M (1986) Direct conversion of cellulose/hemicellulose to ethanol by *Neurospora crassa*. Enzyme Microb Technol 8:149–152. https://doi.org/10.1016/0141-0229(86)90103-1

Eliasson A, Christensson C, Wahlbom CF, Hahn-Hägerdal B (2000) Anaerobic xylose fermentation by recombinant *Saccharomyces cerevisiae* carrying XYL1, XYL2, and XKS1 in mineral medium chemostat cultures. Appl Environ Microbiol 66:3381–3386

Espinosa E, Sánchez R, González Z et al (2017) Rapidly growing vegetables as new sources for lignocellulose nanofibre isolation: physicochemical, thermal and rheological characterisation. Carbohydr Polym 175:27–37. https://doi.org/10.1016/j.carbpol.2017.07.055

Galazka JM, Tian C, Beeson WT et al (2010) Cellodextrin transport in yeast for improved biofuel production. Science 330:84–86. https://doi.org/10.1126/science.1192838

Garcia A (2014) Evaluation of different lignocellulosic raw materials as potential alternative feedstocks in biorefinery processes. ScienceDirect. http://www.sciencedirect.com/science/article/pii/S0926669013006961

Ghose TK, Bandyopadhyay KK (1980) Rapid ethanol fermentation in immobilized yeast cell reactor. Biotechnol Bioeng 22:1489–1496. https://doi.org/10.1002/bit.260220713

Global Bioenergy Partnership, Food and Agriculture Organization of the United Nations, Climate E and TD (2011) The global bioenergy partnership sustainability indicators for bioenergy. FAO, Climate, Energy and Tenure Division, Rome

Gnansounou E, Dauriat A (2010) Techno-economic analysis of lignocellulosic ethanol: a review. Bioresour Technol 101:4980–4991. https://doi.org/10.1016/j.biortech.2010.02.009

Goldemberg J (2007) Ethanol for a sustainable energy future. Science 315:808–810

Gupta R, Sharma KK, Kuhad RC (2009) Separate hydrolysis and fermentation (SHF) of *Prosopis juliflora*, a woody substrate, for the production of cellulosic ethanol by *Saccharomyces cerevisiae* and *Pichia stipitis*-NCIM 3498. Bioresour Technol 100:1214–1220. https://doi.org/10.1016/j.biortech.2008.08.033

Hemansi, Gupta R, Kuhad RC, Saini JK (2018) Cost effective production of complete cellulase system by newly isolated *Aspergillus niger* RCKH-3 for efficient enzymatic saccharification: medium engineering by overall evaluation criteria approach (OEC). Biochem Eng 132:182–190. https://doi.org/10.1016/j.bej.2018.01.019

Hahn-Hägerdal B, Galbe M, Gorwa-Grauslund MF et al (2006) Bio-ethanol—the fuel of tomorrow from the residues of today. Trends Biotechnol 24:549–556. https://doi.org/10.1016/j.tibtech.2006.10.004

Hamelinck CN, van Hooijdonk G, Faaij AP (2005) Ethanol from lignocellulosic biomass: techno-economic performance in short-, middle-and long-term. Biomass Bioenergy 28:384–410. https://doi.org/10.1016/j.biombioe.2004.09.002

Himmel ME, Ding S-Y, Johnson DK et al (2007) Biomass recalcitrance: engineering plants and enzymes for biofuels production. Science 315:804–807. https://doi.org/10.1126/science.1137016

Hong Y, Nizami A-S, Pour Bafrani M et al (2013) Impact of cellulase production on environmental and financial metrics for lignocellulosic ethanol. Biofuels Bioprod Biorefining 7:303–313. https://doi.org/10.1002/bbb.1393

Horn SJ, Vaaje-Kolstad G, Westereng B, Eijsink V (2012) Novel enzymes for the degradation of cellulose. Biotechnol Biofuels 5:45. https://doi.org/10.1186/1754-6834-5-45

Hu CK, Bai FW, An LJ (2005) Effect of flocculence of a self-flocculating yeast on its tolerance to ethanol and the mechanism. Sheng Wu Gong Cheng Xue Bao Chin J Biotechnol 21:123–128

Hu J, Chandra R, Arantes V et al (2015) The addition of accessory enzymes enhances the hydrolytic performance of cellulase enzymes at high solid loadings. Bioresour Technol 186:149–153. https://doi.org/10.1016/j.biortech.2015.03.055

IEA technology (2011) Publication: technology roadmap: biofuels for transport. http://www.iea.org/publications/freepublications/publication/technology-roadmap-biofuels-for-transport.html

Ingram LO, Aldrich HC, Borges ACC et al (1999) Enteric bacterial catalysts for fuel ethanol production. Biotechnol Prog 15:855–866. https://doi.org/10.1021/bp9901062

ISO (2009) ISO 13065:2015—Sustainability criteria for bioenergy. https://www.iso.org/standard/52528.html

ISO (2006) ISO 14040:2006—Environmental management—life cycle assessment—Principles and framework. https://www.iso.org/standard/37456.html

Jeffries TW, Jin Y-S (2004) Metabolic engineering for improved fermentation of pentoses by yeasts. Appl Microbiol Biotechnol 63:495–509. https://doi.org/10.1007/s00253-003-1450-0

Kang Q, Appels L, Tan T, Dewil R (2014) Bioethanol from lignocellulosic biomass: current findings determine research priorities. Sci World J. https://www.hindawi.com/journals/tswj/2014/298153/

Karp A, Shield I (2008) Bioenergy from plants and the sustainable yield challenge. New Phytol 179:15–32. https://doi.org/10.1111/j.1469-8137.2008.02432.x

Kim TH, Taylor F, Hicks KB (2008) Bioethanol production from barley hull using SAA (soaking in aqueous ammonia) pretreatment. Bioresour Technol 99:5694–5702. https://doi.org/10.1016/j.biortech.2007.10.055

Kuhad RC, Gupta R, Khasa YP et al (2011a) Bioethanol production from pentose sugars: current status and future prospects. Renew Sustain Energy Rev 15:4950–4962. https://doi.org/10.1016/j.rser.2011.07.058

Kuhad RC, Gupta R, Singh A (2011b) Microbial cellulases and their industrial applications. Enzyme Res 2011:1–10. https://doi.org/10.4061/2011/280696

Kuhad RC, Mehta G, Gupta R, Sharma KK (2010) Fed batch enzymatic saccharification of newspaper cellulosics improves the sugar content in the hydrolysates and eventually the ethanol fermentation by *Saccharomyces cerevisiae*. Biomass Bioenergy 34:1189–1194. https://doi.org/10.1016/j.biombioe.2010.03.009

Lynd LR (1996) Overview and evaluation of fuel ethanol from cellulosic biomass: technology, economics, the environment, and policy. Annu Rev Environ Resour. http://www.annualreviews.org/doi/abs/10.1146/annurev.energy.21.1.403

Mamma D, Christakopoulos P, Koullas D et al (1995) An alternative approach to the bioconversion of sweet sorghum carbohydrates to ethanol. Biomass Bioenergy 8:99–103. https://doi.org/10.1016/0961-9534(95)00006-S

Matsushika A, Watanabe S, Kodaki T, et al (2008) Bioethanol production from xylose by recombinant *Saccharomyces cerevisiae* expressing xylose reductase, NADP+-dependent xylitol dehydrogenase, and xylulokinase

Millati R, Edebo L, Taherzadeh MJ (2005) Performance of *Rhizopus, Rhizomucor*, and *Mucor* in ethanol production from glucose, xylose, and wood hydrolyzates. Enzyme Microb Technol 36:294–300. https://doi.org/10.1016/j.enzmictec.2004.09.007

Monsalve G, Jhon F, Perez MD et al (2006) Producción De Etanol A Partir De La Cáscara De Banano Y De Almidón De Yuca. Dyna 73:21–27

Morales M, Quintero J, Conejeros R, Aroca G (2015) Life cycle assessment of lignocellulosic bioethanol: environmental impacts and energy balance. Renew Sustain Energy Rev 42:1349–1361. https://doi.org/10.1016/j.rser.2014.10.097

Mountfort DO, Rhodes LL (1991) Anaerobic growth and fermentation characteristics of *Paecilomyces lilacinus* isolated from mullet gut. Appl Environ Microbiol 57:1963–1968

Raud M, Tutt M, Olt J, Kikas T (2016) Dependence of the hydrolysis efficiency on the lignin content in lignocellulosic material. Int J Hydrog Energy 37:16338–16343. https://doi.org/10.1016/j.ijhydene.2016.03.190

Roukas T, Kotzekidou P (1997) Pretreatment of date syrup to increase citric acid production. Enzyme Microb Technol 21:273–276. https://doi.org/10.1016/S0141-0229(97)00041-0

Roy P, Tokuyasu K, Orikasa T et al (2012) A review of life cycle assessment (LCA) of bioethanol from lignocellulosic biomass. Jpn Agric Res Q JARQ 46:41–57. https://doi.org/10.6090/jarq.46.41

RSB (2012) Roundtable on sustainable biofuels recognises rainforest alliance certified farms. https://www.isealalliance.org/online-community/news/roundtable-on-sustainable-biofuels-recognises-rainforest-alliance-certified-farms

Saha BC, Iten LB, Cotta MA, Wu YV (2005) Dilute acid pretreatment, enzymatic saccharification, and fermentation of rice hulls to ethanol. Biotechnol Prog 21:816–822. https://doi.org/10.1021/bp049564n

Saini JK, Saini R, Tewari L (2015) Lignocellulosic agriculture wastes as biomass feedstocks for second-generation bioethanol production: concepts and recent developments. 3 Biotech 5:337–353. https://doi.org/10.1007/s13205-014-0246-5

Sassner P, Galbe M, Zacchi G (2008) Techno-economic evaluation of bioethanol production from three different lignocellulosic materials. Biomass Bioenergy 32:422–430. https://doi.org/10.1016/j.biombioe.2007.10.014

Schneider H, Wang PY, Chan YK, Maleszka R (1981) Conversion of D-xylose into ethanol by the yeast *Pachysolen tannophilus*. Biotechnol Lett 3:89–92. https://doi.org/10.1007/BF00145116

Singh LK, Chaudhary G, Majumder C, Ghosh S (2011) Utilization of hemicellulosic fraction of lignocellulosic biomaterial for bioethanol production. Adv Appl Sci Res 2:508–521

Sonderegger M, Jeppsson M, Larsson C et al (2004) Fermentation performance of engineered and evolved xylose-fermenting *Saccharomyces cerevisiae* strains: xylose-fermenting *S. Cerevisiae* Strains. Biotechnol Bioeng 87:90–98. https://doi.org/10.1002/bit.20094

Sonderegger M, Sauer U (2003) Evolutionary engineering of *Saccharomyces cerevisiae* for anaerobic growth on xylose. Appl Environ Microbiol 69:1990–1998. https://doi.org/10.1128/AEM.69.4.1990-1998.2003

Sreenath HK, Koegel RG, Moldes AB et al (1999) Enzymic saccharification of alfalfa fibre after liquid hot water pretreatment. Process Biochem 35:33–41. https://doi.org/10.1016/S0032-9592(99)00029-1

Sun Y, Cheng J (2002) Hydrolysis of lignocellulosic materials for ethanol production: a review. Bioresour Technol 83:1–11

Tantirungkij M, Nakashima N, Seki T, Yoshida T (1993) Construction of xylose-assimilating *Saccharomyces cerevisiae*. J Ferment Bioeng 75:83–88. https://doi.org/10.1016/0922-338X(93)90214-S

Tenenbaum DJ (2008) Food vs. fuel: diversion of crops could cause more hunger. Environ Health Perspect 116:A254–A257

Thompson P (2012) The agricultural ethics of biofuels: the food vs. Fuel Debate. Agriculture 2:339–358. https://doi.org/10.3390/agriculture2040339

Träff-Bjerre KL, Jeppsson M, Hahn-Hägerdal B, Gorwa-Grauslund M-F (2004) Endogenous NADPH-dependent aldose reductase activity influences product formation during xylose consumption in recombinant *Saccharomyces cerevisiae*. Yeast 21:141–150. https://doi.org/10.1002/yea.1072

US EPA O (2016) Global greenhouse gas emissions data. In: US EPA. https://www.epa.gov/ghgemissions/global-greenhouse-gas-emissions-data

Vaaje-Kolstad G, Westereng B, Horn SJ et al (2010) An oxidative enzyme boosting the enzymatic conversion of recalcitrant polysaccharides. Science 330:219–222. https://doi.org/10.1126/science.1192231

Verho R, Londesborough J, Penttilä M, Richard P (2003) Engineering redox cofactor regeneration for improved pentose fermentation in *Saccharomyces cerevisiae*. Appl Environ Microbiol 69:5892–5897. https://doi.org/10.1128/AEM.69.10.5892-5897.2003

Wyman CE, Hinman ND (1990) Ethanol. Appl Biochem Biotechnol 24–25:735–753. https://doi.org/10.1007/BF02920291

Zhang A-L, Luo J-X, Zhang T-Y et al (2009) Recent advances on the GAP promoter derived expression system of *Pichia pastoris*. Mol Biol Rep 36:1611–1619. https://doi.org/10.1007/s11033-008-9359-4

Zhao J, Xia L (2009) Simultaneous saccharification and fermentation of alkaline-pretreated corn stover to ethanol using a recombinant yeast strain. Fuel Process Technol 90:1193–1197. https://doi.org/10.1016/j.fuproc.2009.05.018

Bioethanol Production Using *Saccharomyces cerevisiae* Immobilized in Calcium Alginate–Magnetite Beads and Application of Response Surface Methodology to Optimize Bioethanol Yield

Snehal Ingale, Venkata Anand Parnandi and Sanket J. Joshi

Abstract We studied the bioethanol production in molasses-based medium by yeast *Saccharomyces cerevisiae* immobilized in calcium alginate magnetite beads (CAMB). The yeast was isolated from soil samples collected near a local sugar mill, and identified as *S. cerevisiae*. We synthesized magnetite nanoparticles and immobilized yeast in CAMB. The media components and environmental parameters were statistically screened and optimized for better ethanol production, using statistical design methodologies—factorial designs and response surface methodology. The factors of molasses concentration, temperature and incubation time were found to have significant effect on ethanol production. The immobilized cells could be reused for more than 120 days, retaining its original activity. The CAMBs with immobilized yeast cells were analysed by ESEM with EDAX, after 96 h of fermentation to observe the surface structure of the beads. It can be observed that yeast was immobilized in the beads and actively growing. Further ethanol production was carried out in packed-bed column reactor using yeast immobilized in CAMB, under fed-batch mode. The average ethanol produced by fed-batch fermentation was 1.832 g% $\pm$ 0.103, and the average ethanol yield was 81.420% $\pm$ 4.6. Further studies using yeast immobilized in CAMB are recommended to carry out continuous fermentation, and further scale up

S. Ingale · V. A. Parnandi
Ashok and Rita Patel Institute of Integrated Study& Research in Biotechnology and Allied Sciences (ARIBAS), Sardar Patel University, New Vallabh Vidhyanagar, Anand, Gujarat, India

S. J. Joshi (✉)
Central Analytical and Applied Research Unit, College of Science, Sultan Qaboos University, Muscat, Oman
e-mail: sanketjj@gmail.com; sanket@squ.edu.om

N. Srivastava et al. (eds.), *Sustainable Approaches for Biofuels Production Technologies*, Biofuel and Biorefinery Technologies 7,
https://doi.org/10.1007/978-3-319-94797-6_9

bioethanol production in a magnetically stabilized fluidized bed reactor (MSFBR), where the position of the beads in the system can be controlled and maintained by the application of oscillating electric field.

1 Introduction

Nanotechnology and biofuels are two research fields which are exponentially growing. In past few years, nanoparticles have been found to be useful in various applications like in electronics, as catalysts and as an antimicrobial agent, photocatalytic degradation of organic dyes, in enhancing oil recovery, in health and environmental applications, to list few (Roy et al. 2014; Vanaja et al. 2014; Muthukrishnan et al. 2015). Generally, nanoparticles are chemically synthesized using processes involving reducing agents and capping agents under controlled conditions, or green synthesis using microorganisms or plant-based products (Abdul Rahman et al. 2014; Priyadarshini et al. 2014; Padman et al. 2014; Singh et al. 2015). The world economy has been dominated by technologies that depend solely on fossil energies, such as natural gas, coal or petroleum to produce chemicals, fuels, materials and power. A 50% rise in worldwide marketed energy expenditure has been projected by the US Energy Information Administration between 2005 and 2030. This growth will be predominantly observed in the non-OECD (Organization for Economic Co-operation and Development) or developing world countries where energy consumption is expected to increase by 85%, corresponding to the data collected by EIA in 2008 about 40.1% of world consumption (Mino 2010). Energy security and environmental concerns are largely the reason behind the growth of biofuels around the globe. To facilitate their growth, a wide range of incentives, market mechanisms and subsidies have been put in place. Biofuels provide an alternative to fossil fuel dependency and emit fewer pollutants (De Carvalho et al. 1993). Apart from these considerations, underdeveloped countries also view biofuels as a potential means to create employment opportunities as well as stimulate rural development. For example, India ranks sixth in terms of energy demand, accounting for 3.6% of the total global energy demand. Crude oil has been the major resource to meet the energy demand, and the demand for oil and its products is increasing dramatically every year. In India, biofuels are based mainly on feedstocks which are non-food based, to avoid a possible conflict of fuel versus food security. By year 2025, most of the petroleum in India will be imported. Estimates have indicated that more than 150 million tonnes of crude oil was consumed in 2007–08. The domestic crude oil is only able to meet around 23% of the actual demand, while the rest of the demand is fulfilled by importing crude oil from other countries. In 2008, in an effort to increase its energy security and independence, the National Policy on Biofuels was announced by the Government of India, mandating a phase-wise implementation of the programme of ethanol blended in petrol in various states. The oil marketing companies (OMCs) were to take up the blending of ethanol at 5% with petrol in 20 states and four national territories. However, due to shortage in ethanol, the implementation of this policy has not had much success (Ray et al. 2012). Bioethanol is an eco-friendly alternate biofuel that can be used in unmodified petrol engines with current fuelling infrastructure and it is easily applicable in the present-day combustion engine, as mixing with gasoline (Hansen et al. 2005). Relatively low emission of carbon monoxide, oxides of nitrogen and other volatile organic compounds are the product of ethanol combustion. Emission from ethanol combustion is lower compared to the emission of fossil fuel combustions such as diesel and gasoline, and its toxicity is also low (Wyman and Hinman 1990).

2 Substrates Used for Bioethanol Production

Bioethanol can be produced from sugar, biomass and wastes. However, the nature of the substrate greatly affects the processes of the ethanol fermentation. Therefore, the raw materials selected for ethanol fermentation have great importance in the fermentation process (Baptista et al. 2006). Hydrolysed enzymes ferment the complex sugars to reducing sugars and then to high concentrations of ethanol. It is also being made from a variety of agricultural by-products such as grain, fruit juices, fruit extracts, whey, sulphite waste liquor and molasses (Nigam et al. 1998). Generally, molasses is extracted from different agricultural sources such as sugarcane and sugar beet. It is a sugary–syrupy dark material left after the extraction of sugar from the mother syrup, and it is very rich in nutrients required by most microorganisms. Molasses are generally found to contain 45–60% total sugars, 20–25% reducing sugars, 25–35% sucrose, 10–16% ash, 0.4–0.8% calcium, 0.1–0.4% sodium, 1.5–5% potassium and pH 5–5.5 (Chen and Chou 1993). Molasses has no furfural, which is toxic to most of fermenting microorganisms (El-Gendy et al. 2013). Generally, cane molasses is reported to contain less sucrose and more invert sugar, and lower nitrogen and raffinose, dark colour and extra buffer capacity (Wang et al. 1984; Borzani et al. 1993, Borzani 2001). Although Brazil produces the most sugarcane, India is the world's largest producer of sugar. The majority of the sugarcane grown in India is used by sugar mills to produce sugar and its main by-products: molasses and bagasse. Currently, 70% of the harvested sugarcane is utilized by regulated mills to produce sugar. The other 20-30% is used for the production of alternate sweeteners: gur and khandsari (Jaggery) and for seeds (Raju et al. 2009).

3 *Saccharomyces Cerevisiae* for Bioethanol Production

Worldwide demand of ethanol is generally satisfied by biotechnological fermentation process but various processes have been developed for ethanol production. Screening of a number of organisms for ethanol production has been performed, which include fungi, yeast and bacteria. These organisms have been studied extensively to determine their ethanol fermentation capabilities, especially yeast cells (Bajaj et al.

2001). *S. cerevisiae* is one such highly studied and utilized eukaryotic microorganism—yeast is a unicellular microorganism. *S. cerevisiae* cells measure 5–10 microns wide and 5–12 μm long. *S. cerevisiae* was originally believed to have been isolated from the skin of grapes (Pretorius 2000). It has an optimum temperature growth range at 30 °C, and it is tolerant of a wide pH range (2.4–8.2), being the optimum pH for growth between values of 3.5 and 3.8 (Gray 1941, 1948). With respect to the nutritional requirements, all strains can grow aerobically on glucose, fructose, sucrose and maltose and fail to grow on lactose and cellobiose. Also, all strains of *S. cerevisiae* can use ammonia and urea as the sole nitrogen source but cannot use nitrate since they lack the ability to reduce them to ammonium ions. They can also use most amino acids, small peptides and nitrogen bases as a nitrogen source (Bisson 1999). Ethanol is produced by fermentation when certain species of yeast (notably *S. cerevisiae*) metabolize sugar in the absence of oxygen, producing ethanol and carbon dioxide. Ethanol is well known as an inhibitor of growth of microorganisms. It has been reported to damage mitochondrial DNA in yeast cells (Ibeas and Jimenez 1997) and to cause inactivation of some enzymes. Nevertheless, some strains of the yeast *S. cerevisiae* show tolerance and can adapt to high concentrations of ethanol (Ghareib et al. 1988; Alexandre et al. 1994). Many studies have documented the alteration of cellular lipid composition in response to ethanol exposure (Mishra and Prasad 1989; Ingram 1976). It has been found that *S. cerevisiae* cells grown in the presence of ethanol appear to increase the amount of monounsaturated fatty acids in cellular lipids (Beaven et al. 1982).

S. cerevisiae can be used as either free cells or as immobilized to different matrices for ethanol production. Immobilization is a general term describing a wide variety of the cell or particle attachment or entrapment (López et al. 1997). It can be applied to basically all types of biocatalysts including enzymes, cellular organelles, animal cells and plant cells. The major advantage of immobilized cells, in contrast to free-living cells and immobilized enzymes, is reduction of the cost of bioprocessing as there is no involvement of pure enzymes, which are very costly even when procured in small quantities, and no requirement for additional steps of cell separation. The biocatalyst can be used repeatedly and continuously, and high cell density is maintained. In addition, immobilization can provide resistance to shear for shear-sensitive cells such as those from plants and animals. Different immobilization types have been defined: covalent coupling/cross-linking, capture behind semipermeable membrane or encapsulation, entrapment and adsorption (Mallick 2002). The types of immobilization can be grouped as 'passive' (using the natural tendency of microorganisms to attach to surfaces—natural or synthetic, and grow on them) and 'active' (flocculent agents, chemical attachment and gel encapsulation) (Cassidy et al. 1996; Cohen 2001; Moreno-Garrido 2008). The use of calcium alginate for immobilization of yeast cells has been around since 1980s. Calcium alginate is preferred because beads made of alginate can be stable for a period of more than 90 days (Nagashima et al. 1983). Cells immobilized on a variety immobilization matrix show comparatively higher yield when utilized for ethanol production (Black et al. 1984; McGhee et al. 1984) as compared to free-living cells. A variety of supports for the immobilization of *S. cerevisiae* were studied, such as spheres of stainless steel (Black et al. 1984), cellulose (Okita et al. 1985), calcium alginate (McGhee et al. 1984), synthetic commercial sponge (Del Borghi et al. 1985) cotton cloth (Joshi and Yamazaki 1984), immobilized cell reactor (Najafpour et al. 2004), yeast anchored on calcium alginate and clay support (Osawemwenze and Adogbo 2013). *S. cerevisiae* cells were entrapped in a matrix of alginate and magnetic nanoparticles (CAMB) and covalently immobilized on magnetite-containing chitosan (CHMM) and cellulose-coated magnetic nanoparticles (CCMN) (Ivanova et al. 2011).

4 Magnetite Nanoparticles

Magnetite (Fe_3O_4) is a biocompatible material, with low toxicity and strong magnetic properties, which responds to an external magnetic field, but not interacting in the absence of magnetic field. It has (Huang et al. 2003). It has been widely used for in vivo examination including magnetic resonance imaging, contrast enhancement, tissue-specific release of therapeutic agents, gene therapy (Berry and Curtis 2003), hyperthermia (Tartaj et al. 2006) and magnetic field assisted radionuclide therapy (Pankhurst et al. 2003), as well as in vitro binding of proteins and enzymes.

It also has biological and medical applications which include tissue repair, immunoassay, detoxification of biological fluids and cell separation (Gupta and Gupta 2005). Due to ferromagnetic properties of magnetite and diamagnetic properties of accompanying molecules and particulate matter, loaded magnetic adsorbents and carriers can be separated from suspensions with the use of magnetic fields (Šafařík and Šafaříková 1999, 2001, 2002). There are a variety of methods for the synthesis of magnetite nanoparticles in various irregular shapes, such as co-precipitation, ultrasound irradiation, hydrothermal and electrochemical synthesis, and pyrolysis, which produce nanoparticles with sizes ranging from $\approx$5 to 100 nm (Nyirő-Kósa et al. 2012; El Ghandoor et al. 2012).

5 Case Study

We studied immobilization of locally isolated *S. cerevisiae* yeast strain in calcium alginate magnetite beads (CAMB), to produce ethanol. The media components and environmental parameters were statistically screened and optimized for better ethanol production, using statistical design methodologies. Further, ethanol production was carried out in packed-bed column reactor using yeast immobilized in CAMB, under fed-batch mode.

5.1 *Materials and Chemicals*

All the chemicals were of analytical grade, purchased from Loba Chemie, HiMedia, Baroda Chemical Industries Ltd., SuLab, and Fisher Scientific, India Ltd. Rose bengal chloramphenicol (RBC) HiVeg agar was purchased from HiMedia, India. Sugarcane molasses and sugarcane bagasse were procured from local farmers and market. When not in use, the molasses was stored at 4 °C.

5.2 *Isolation and Maintenance of Yeast*

The yeast *S. cerevisiae* was isolated using RBC HiVeg agar medium from soil samples collected near a local sugar mill. The culture was maintained on RBC HiVeg agar medium by sub-culturing every 15 days and incubating at 30 °C for 24 h. The yeast culture was also preserved in 25% glycerol solutions for long-term preservation at 5 °C.

5.3 *Synthesis of Magnetite (Fe_3O_4) Nanoparticles*

Ultra-fine particles of magnetite nanoparticles (Fe_3O_4) were prepared by co-precipitating aqueous solutions of $(NH_4)_2Fe(SO_4)_2$ and $FeCl_3$ mixtures, respectively, in alkaline medium. $(NH_4)_2Fe(SO_4)_2$ and $FeCl_3$ solutions were mixed in their respective stoichiometry (i.e. ratio Fe^{+2}: Fe^{+3} $= 1{:}2$). The mixture was kept at 80 °C. This mixture was added to the boiling solution of NaOH (0.5 mol. was dissolved in 600 mL of distilled water) within 10 s under constant stirring. Magnetite was formed by conversion of metal salts into hydroxides, which took place immediately, and transformation of hydroxides into ferrites. The solution was maintained at 100 °C for 1.5 h. The Fe_3O_4 particles were washed several times by distilled water (El Ghandoor et al. 2012) and dried in an oven.

5.4 *Immobilization of* S. Cerevisiae

Active cultures of *S. cerevisiae* for fermentation were prepared in Wickerham WH media: 2 g/L, KH_2PO_4); 10 g/L, $(NH_4)_2SO_4$; 1 g/L, $MgSO_4{\cdot}7H_2O$; 2 g/L, Yeast extract; and 10 g/L, Glucose (Haynes et al. 1955) for 48 h at 30 °C. Cells were harvested by centrifugation and were washed twice with sterile saline (0.85 g NaCl in 100 mL distilled water) and then suspended in sterile saline to be used later for inoculation. The *S. cerevisiae* cells were harvested at exponential phase to be utilized for immobilization. Sodium alginate (2%) was completely dissolved over a period of 4 h by continuous stirring on magnetic stirrer and then autoclaved at 121 °C for 15 min. The cells were mixed with the sodium alginate slurry. The final inoculum size was 10 mg of cell dry weight/mL of gel, and the final content of nanoparticles was 2.5% (w/v). The contents were thoroughly mixed for even dispersal of cells as well as magnetite nanoparticles. To prepare the beads, the slurry, containing yeast cells and magnetite nanoparticles, was dispersed dropwise using a sterile syringe and plunger into chilled 2% $CaCl_2$ solution which was previously sterilized. As soon as the drops of sodium alginate came in contact with the chilled calcium chloride solution, spherical beads were formed as the sodium ion was replaced by calcium ions. The calcium alginate magnetite beads (CAMB) containing the cells were incubated at 4 °C for proper chelation and then thoroughly washed with distilled water repeatedly. The beads were then placed in fresh sterile $CaCl_2$ solution and refrigerated to be used for further studies.

5.5 *Analytical Methods*

Reducing sugars were analysed by dinitrosalicylic acid (DNS) method (Miller 1959). The reducing sugar concentration in the sample was calculated using the standard curve of D-glucose. Total soluble carbohydrate in molasses was determined by the phenol–sulfuric acid method (Dubois et al. 1956; Joshi et al. 2008; Al-Bahry et al. 2013), and concentration of carbohydrates was estimated by comparing it with standard sucrose and glucose solutions. The ethanol concentration was determined by dichromate oxidation and thiosulphate titration (Marcelle et al. 2007; Ingale et al. 2014).

5.6 *Experimental Designs*

5.6.1 Screening Design

Screening was carried out by Plackett–Burman design of the most important components affecting bioethanol production by *S. cerevisiae* using sugarcane molasses. For the designing of experiments, Design Expert software 9.0.4.1 (Stat-Ease, Inc., Minneapolis, MN, USA) was utilized. A total of 11 components were evaluated, with each being represented at two levels, High (H) and Low (L). In the design, it is assumed that the main factors have individual effects but no interactions, and a first-order polynomial equation is appropriate (Eq. 1):

$$Y = \beta_0 + \sum_{i=1}^{n} \beta_i x_i \tag{1}$$

Table 1 Variables showing fermentation parameters used in Plackett–Burman design

Variable	Medium Component	H (+)	L (−)
A	Molasses	20	10
B	Potassium di-hydrogen phosphate	0.5	0.1
C	Ammonium sulphate	2	0.2
D	Magnesium sulphate	0.2	0.05
E	Yeast extract	0.50	0.10
F	pH	7.00	5.00
G	Temperature	37	28
H	Incubation	96	48
J	Immobilized yeasts	10	5
K	Agitation	100	0
L	Pretreated hydrolysate	5	1

where Y represents the response, β_0 is the model coefficient, β_i is the linear coefficient, x_i is the variables and n is the number of parameters (variables). The effect of each variable was determined by Eq. 2:

$$E_{(Xi)} = \frac{\sum M_{i+} - \sum M_{i-}}{N} \tag{2}$$

where $E_{(xi)}$ is the response value effect of the tested variable, ΣM_{i+} is the summation of the response value at high level, ΣM_{i-} is the summation of the response value at low level and N is the number of experiments. Table 1 represents the factors to be evaluated. Table 2 shows the design matrix built by statistical software Design Expert software 9.0.4.1 (Stat-Ease, Inc., Minneapolis, MN, USA) for the evaluation of 11 variables in 20 experimental runs. Variables A through L represent the 11 medium components (actual variables) and D1 through D8 represent dummy variables (used to reduce error in data). Data were analysed through analysis of variance (ANOVA).

5.6.2 Optimization Design

After selecting the most important variables which influenced the bioethanol production by *S. cerevisiae*, response surface methodology (RSM) was used for optimization of the process. The central composite design (CCD) was applied to study the different process variables. The behaviour of the system was demonstrated by the following quadratic equation (Eq. 3):

$$Y = \beta_0 + \sum \beta_i x_i + \sum \beta_{ij} x_i x_j + \sum \beta_{ii} x_i^2 \tag{3}$$

Table 2 Plackett–Burman design for 19 factors

Trial	A	B	C	D	E	F	G	H	J	K	L	D1	D2	D3	D4	D5	D6	D7	D8	Ethanol (g%)	%Yield
1	H	H	L	L	H	H	H	H	L	H	L	H	L	L	L	L	H	H	L	1.28	43.38
2	L	H	H	L	L	H	H	H	H	L	H	L	H	L	L	L	L	H	H	1.07	72.29
3	H	L	H	H	L	L	H	H	H	H	L	H	L	H	L	L	L	L	H	1.35	45.58
4	H	H	L	H	H	L	L	H	H	H	H	L	H	L	H	L	L	L	L	1.45	49.15
5	L	H	H	L	H	H	L	L	H	H	H	H	L	H	L	H	L	L	L	0.97	65.27
6	L	L	H	H	L	H	H	L	L	H	H	H	H	L	H	L	H	L	L	1.48	100.3
7	L	L	L	H	H	L	H	H	L	L	H	H	H	H	L	H	L	H	L	0.88	59.29
8	L	L	L	L	H	H	L	H	H	L	L	H	H	H	H	L	H	L	H	1.36	92.18
9	H	L	L	L	L	H	H	L	H	H	L	L	H	H	H	H	L	H	L	1.2	40.41
10	L	H	L	L	L	L	H	H	L	H	H	L	L	H	H	H	H	L	H	1.24	83.91
11	H	L	H	L	L	L	L	H	H	L	H	H	L	L	H	H	H	H	L	1.14	38.38
12	L	H	L	H	L	L	L	L	H	H	L	H	H	L	L	H	H	H	H	0.79	53.28
13	H	L	H	L	H	L	L	L	L	H	H	L	H	H	L	L	H	H	H	1.4	47.29
14	H	H	L	H	L	H	L	L	L	L	H	H	L	H	H	L	L	H	H	1.41	47.75
15	H	H	H	L	H	L	H	L	L	L	L	H	H	L	H	H	L	L	H	1.9	64.36
16	H	H	H	H	L	H	L	H	L	L	L	L	H	H	L	H	H	L	L	1.13	38.07
17	L	H	H	H	H	L	H	L	H	L	L	L	L	H	H	L	H	H	L	1.11	75.21
18	L	L	H	H	H	H	L	H	L	H	L	L	L	L	H	H	L	H	H	0.79	53.4
19	H	L	L	H	H	H	H	L	H	L	H	L	L	L	L	H	H	L	H	1.63	55.1
20	L	L	L	L	L	L	L	L	L	L	L	L	L	L	L	L	L	L	L	1.51	101.9

where Y is the predicted response, β_0 is a constant, β_i is the linear coefficient, β_{ii} is the squared coefficient, β_{ij} is the cross-product coefficient and x_i is the dimensionless coded value of (X_i). The above equation was solved by using the statistical software Design Expert software 9.0.4.1 (Stat-Ease, Inc., Minneapolis, MN, USA). A 2^5 factorial design with five replicates at the centre point with a total number of 20 trials were employed.

5.7 Packed-Bed Fermentation Under Fed-Batch Mode

After optimization of bioethanol production by batch fermentation, fed-batch bioethanol fermentations were also carried out. The diameter of the column was 3 cm, and the length of the column was 45 cm. The volume of the column without beads was 230 mL. When the column was packed with CAMB, the void volume of the column was 100 mL. The column was packed to 70% of the column volume. The reactor was set up using standard IV (Intravenous) infusion set to control the feed rate (Fig. 1). The fed-batch fermentation was carried out at molasses concentration of 20 g% (w/v), temperature of 28 °C and incubation time of 72 h. The feeding rate was 0.06 g mL^{-1} h^{-1}.

5.8 Scanning Electron Microscopy of CAMB

For electron microscope scanning, samples of calcium alginate magnetite beads (CAMB) immobilized with *S. cerevisiae* were taken after 96 h of ethanol fermentation. The samples were examined under a scanning electron microscope model-XL 30 ESEM with EDAX (Philips, Netherlands). The resolution of the instrument was up to 2 Å, acceleration voltage of 0.2–30 kV and up to 2,50,000× magnification. The analysis was performed at Charutar Vidya Mandal's SICART (Sophisticated Instrumentation Centre for Applied Research and Testing) facility, Gujarat, India.

6 Results and Discussions

6.1 Isolation of Yeast and Immobilization in Calcium Alginate Magnetite Nanoparticles

The yeast was locally isolated and identified as *S. cerevisiae* (Fig. 2). The synthesized and dried magnetite nanoparticles used to prepare calcium alginate beads are shown in Fig. 3. *S. cerevisiae* was immobilized in calcium alginate magnetite beads as described and was stored in fridge prior to the experiment (Fig. 4). The concentra-

Fig. 1 Column reactor for bioethanol fermentation

tions of total sugars as well as reducing sugar present in molasses were analysed by different methods. The concentration of total sugar present in cane molasses analysed in the present study was 29.42%, which was more or less similar to that reported by others (Nofemele et al. 2012; Bajaj et al. 2003). Sugarcane bagasse was pretreated with 5% H_2SO_4 for 2 h, was neutralized with NaOH and was used as a pretreated hydrolysate in the optimization experiments to see its effect on ethanol production. There was no requisite for pretreatment of the molasses as it does not contain complex compounds such as cellulose and lignin.

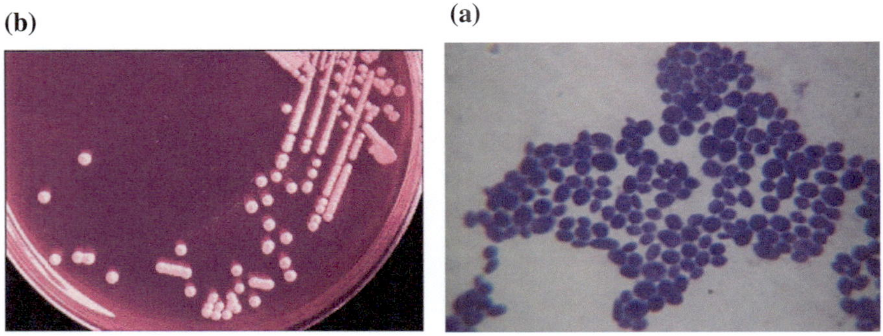

Fig. 2 Rose bengal agar plate and *S. cerevisiae* on RBA plate and stained with crystal violet

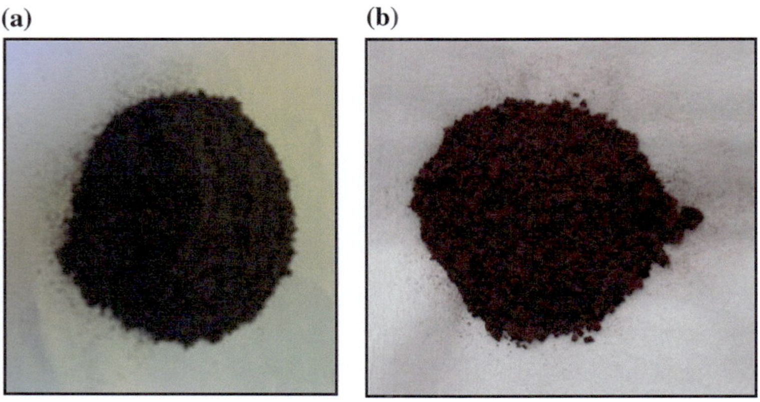

Fig. 3 Magnetite nanoparticles powder (**a**, **b**)

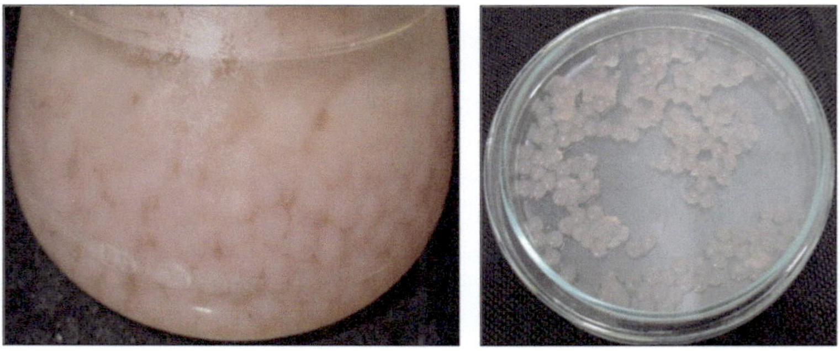

Fig. 4 Yeast cells immobilized in calcium alginate magnetite beads (CAMB)

6.2 Screening of Variables by Plackett–Burman Design

The statistical design used for the optimization of ethanol production was an 11-factor system with eight dummy variables, with the factors being molasses, potassium di-hydrogen phosphate, ammonium sulphate, magnesium sulphate, yeast extract, pH, temperature, incubation time, agitation and pretreated sugarcane bagasse hydrolysate. The responses of the system were the ethanol production and ethanol yield. The design summary is shown in Table 3.

The design was used to identify the most important factors early in the experimentation phase in order to screen out the factors which have significant impact on bioethanol production as compared to other less significant factors. It was observed that runs 6, 8, 10, 17 and 20 had maximum ethanol production and maximum ethanol yield.

The adequacy of the factorial model for the experimental responses (ethanol production R1 and yield R2) was checked using the analysis of variance (ANOVA), which was verified using the Fisher's statistical model (F-value). Table 4 shows the ANOVA for R1 response. ANOVA of the factorial model for ethanol production had the 'Model *F-value*' of 3.24, which implied the model was significant. There was only a 4.98% chance that a 'Model F-value' this large could occur due to noise.

Since '*p-value*' of the model was less than 0.0500, it indicated that the model was significant. The '*p-value*' of molasses (A) was 0.004, which is less than 0.0500, which indicated that the factor had a significant effect on the production of ethanol.

A normal probability of the standardized residuals for ethanol production is shown in Fig. 5. A normal probability plot indicates that if the residuals follow a normal distribution, in which case, the points will follow a straight line. Since some scattering is expected even with the normal data, it can be assumed that the data is normally distributed. Thus, it indicates a good validity for the approximation of factorial model.

Table 5 shows the ANOVA for R1 response. ANOVA of the factorial model for bioethanol yield had the 'Model *F-value*' of 7.64, which implied the model was significant. There was only a 0.22% chance that a 'Model F-value' this large could occur due to noise. Since '*p-value*' of the model was less than 0.0500, it indicated that the model was significant. The '*p-value*' of molasses (A) was 0.0022, which is less than 0.0500, which indicated that the factor had a significant effect on the yield of ethanol.

A normal probability of the standardized residuals for ethanol production is shown in Fig. 6. A normal probability plot indicates that if the residuals follow a normal distribution, in which case, the points will follow a straight line. Since some scattering is expected even with the normal data, as shown in Fig. 6, it can be assumed that the data is normally distributed. Thus, the obtained probability plot indicates a good validity for the approximation of the factorial model. Based on the results obtained from the Plackett–Burman design, we selected three variables, namely, molasses concentration, temperature and the incubation time. Molasses concentration, temperature and the incubation time have positive influence on bioethanol production and yield hence higher levels of all the three variables resulted in higher bioethanol

Table 3 Plackett–Burman factorial design summary for optimization of ethanol production

Factor	Name	Units	Min	Max	Low	High	Mean	St. Dv.
A	Molasses	g% (w/v)	10	20	−1	1	15	5.13
B	Potassium di-hydrogen phosphate	g%	0.1	0.5	−1	1	0.3	0.21
C	Ammonium sulphate	g%	0.2	2	−1	1	1.1	0.92
D	Magnesium sulphate	g%	0.05	0.2	−1	1	0.125	0.08
E	Yeast extract	g%	0.1	0.5	−1	1	0.3	0.21
F	pH	pH	5	7	−1	1	6	1.03
G	Temperature	C	28	37	−1	1	32.5	4.62
H	Incubation time	Hrs	48	96	−1	1	72	24.62
J	Immobilized yeast	%(w/v)	5	10	−1	1	7.5	2.57
K	Agitation	Rpm	0	100	−1	1	50	51.30
L	Pretreated hydrolysate	%(v/v)	1	5	−1	1	3	2.052

Resp.	Name	Unit	Obs.	Anal.	Min	Max	Mean	Std. Dev.	Ratio	Trans
R1	Ethanol	g%	20	Factorial	0.79	1.90	1.25	0.28	2.42	None
R2	Yield	g%	20	Factorial	38.07	101.94	61.33	20.19	2.68	None

Table 4 Analysis of variance table of the factorial model (Plackett–Burman) for bioethanol production

Source	ΣH	ΣL	Diff.	Eff.	Mean square	Mean square for error	t	p-value	95% CI
A-molasses	13.89	11.20	2.69	0.27	0.072	0.018	3.97	0.004	99.59
B-KH_2PO_4	12.35	12.74	-0.39	-0.04	0.002		0.08	0.935	6.45
C-$(NH_4)_2SO_4$	12.34	12.75	-0.41	-0.04	0.002		0.09	0.929	7.12
D-$MgSO_4$	12.02	13.07	-1.05	-0.11	0.011		0.61	0.562	43.83
E-yeast extract	12.77	12.32	0.45	0.05	0.002		0.11	0.914	8.58
F-pH	12.32	12.77	-0.45	-0.05	0.002		0.11	0.914	8.58
G-temperature	13.14	11.95	1.19	0.12	0.014		0.78	0.459	54.08
H-incubation time	13.40	11.69	1.71	0.17	0.029		1.61	0.147	85.30
J-immobilized yeast	12.07	13.02	-0.95	-0.10	0.009		0.50	0.634	36.65
K-agitation	11.95	13.14	-1.19	-0.12	0.014		0.78	0.459	54.08
L-pretreated hydrolysate	12.67	12.42	0.25	0.03	0.001		0.03	0.97	2.65

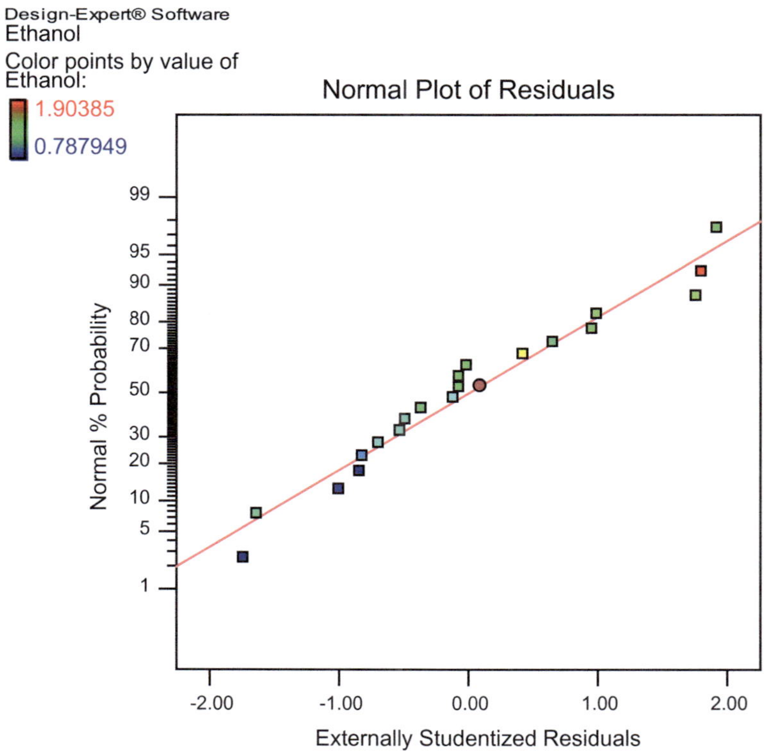

Fig. 5 Normal probability plots of the residuals for bioethanol production

production. The other components of the production medium, such as KH_2PO_4, $(NH_4)_2SO_4$, $MgSO_4.7H_2O$, yeast extract, pH, pretreated hydrolysate, immobilized yeast and agitation, were found to be insignificant, so their concentrations were set at their middle level in central composite design.

6.3 *Optimization of Bioethanol Production and Yield by Response Surface Methodology (RSM)*

6.3.1 Statistical Analysis and Validation of Model

The statistical design used for the bioethanol production is a three-factor (Molasses concentration, temperature and the incubation time) system. A total of 20 experiments with three variables and five coded levels (five different concentrations) were performed. Table 6 shows the coded and actual values of the variables. The response of

Table 5 Analysis of variance table of the factorial model (Plackett–Burman) for bioethanol yield

Source	ΣH	ΣL	Diff.	Effect	Mean square	Mean square for error	t	p-value	95% CI
A-molasses	757.1	469.5	287.6	28.8	827.1	61.0	13.6	0.0	100.0
B-KH_2PO_4	592.7	633.9	−41.2	−4.1	17.0		0.3	0.8	21.2
C-$(NH_4)_2SO_4$	600.1	626.4	−26.2	−2.6	6.9		0.1	0.9	8.7
D-$MgSO_4$	577.1	649.4	−72.3	−7.2	52.3		0.9	0.4	58.3
E-yeast extract	604.6	621.9	−17.3	−1.7	3.0		0.0	1.0	3.8
F-pH	608.1	618.4	−10.3	−1.0	1.1		0.0	1.0	1.3
G-temperature	639.8	586.7	53.1	5.3	28.2		0.5	0.7	34.4
H-incubation time	650.9	575.6	75.3	7.5	56.7		0.9	0.4	62.0
J-immobilized yeast	586.9	639.7	−52.8	−5.3	27.9		0.5	0.7	34.1
K-agitation	582.0	644.6	−62.6	−6.3	39.2		0.6	0.5	46.1
L-pretreated hydrolysate	618.7	607.8	10.9	1.1	1.2		0.0	1.0	1.5

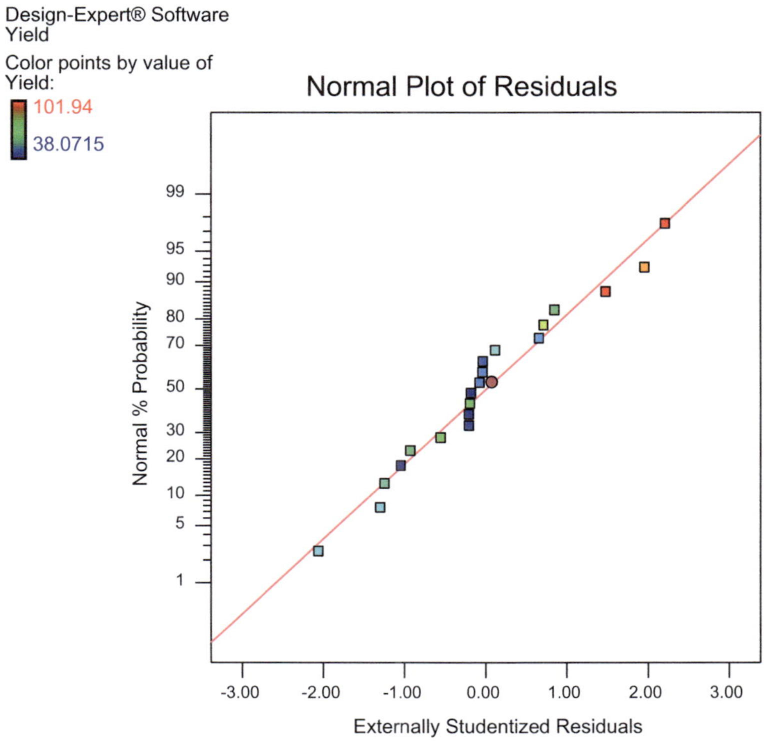

Fig. 6 Normal probability plots of the residuals for bioethanol yield

Table 6 Coded and actual values of the variables used in central composite design

Independent variables	Level				
	−α	−1	0	1	α
Molasses	6.59	10	15	20	23.41
Temperature	24.93	28	32.5	37	40.07
Incubation time	31.64	48	72	96	112.36

the production was based on the ethanol and yield. The design summary is shown in Table 7.

The design was a set of 20 runs, combinations of three-factor experimental design, based on the RSM and CCD as shown in Table 8. The RSM is a mathematical-based system utilized to study the interactions between the factors, while the CCD enables the deduction of optimal condition for bioethanol production. CCD contains an embedded factorial or fractural factorial design with centre points that is augmented with a group of 'star points' that allow estimation of curvature. As shown in Table 8, runs 12, 13, 17 and 20 had maximum bioethanol production and maximum

Table 7 Quadratic model, response surface design summary for optimization of bioethanol production

Factor	Name	Units	Type	Subtype	Min	Max	Coded	Values	Mean	Std. Dev.
A	Molasses	g%(w/v)	Num.	Cont.	6.591	23.41	−1 = 10	1 = 20	15	4.22
B	Temperature	C	Num.	Cont.	24.93	40.07	−1 = 28	1 = 37	32.5	3.82
C	Incubation time	Hrs	Num.	Cont.	31.64	112.36	−1 = 48	1 = 96	72	20.35
Resp.	Name	Units	Obs.	Analysis	Min	Max	Mean	St. Dv.	Ratio	Trans
R1	Ethanol	g%	20	Polynom.	0.56	2.55	1.43	0.52	4.57	None
R2	Yield	%	20	Polynom.	35.42	85.97	63.93	15.35	2.43	None

Table 8 Test design and results of response surface analysis

Std	Run	Factor1 A: Molasses g % (w/v)	Factor2 B: Temperature (°C)	Factor3 C: Incubation time (h)	Response1: Ethanol	Response2: Yield
14	1	15	32.5	112.36	1.766	78.477
2	2	20	28	48	1.496	49.856
17	3	15	32.5	72	1.213	53.907
11	4	15	24.93	72	1.318	58.581
5	5	10	28	96	1.131	75.373
13	6	15	32.5	31.64	0.797	35.419
1	7	10	28	48	0.736	49.060
18	8	15	32.5	72	1.361	60.474
12	9	15	40.07	72	1.591	70.710
10	10	23.41	32.5	72	2.164	61.623
9	11	6.59	32.5	72	0.557	56.312
6	12	20	28	96	2.548	84.894
16	13	15	32.5	72	1.819	80.843
4	14	20	37	48	1.099	36.630
3	15	10	37	48	0.760	50.635
15	16	15	32.5	72	1.473	65.436
20	17	15	32.5	72	1.935	85.967
8	18	20	37	96	1.935	64.475
19	19	15	32.5	72	1.764	78.385
7	20	10	37	96	1.222	81.466

yield. The quadratic polynomial equations describe the correlation between the significant coefficients, i.e. p-value (Prob > F) less than 0.05, and are used to obtain the regression values of coefficients where only significant coefficients are considered. Since this model supports hierarchy, the insignificant coefficients were not omitted. This equation was used to derive the predicted responses for ethanol (Eq. 4) and yield (Eq. 5).

$$\begin{aligned}\textbf{Ethanol} = {} & 1.59379 + 0.43436 \times A - 0.03188 \times B + 0.32029 \times C \\ & - 0.14058 \times A \times B + 0.12870 \times A \times C - 0.01851 \times B \\ & \times C - 0.07964 \times A^2 - 0.04631 \times B^2 - 0.10756 \times C^2 \end{aligned} \quad (4)$$

$$\begin{aligned}\textbf{Yield} = {} & 70.77055 - 0.85999 \times A - 0.40831 \times B + 14.09129 \\ & \times C - 5.16414 \times A \times B + 0.71733 \times A \times C - 0.33432 \\ & \times B \times C - 3.77229 \times A^2 - 1.76472 \times B^2 - 4.48627C^2 \end{aligned} \quad (5)$$

The adequacy of the quadratic model for the experimental responses (Ethanol R1 and Yield R2) was checked using the analysis of variance (ANOVA), which was

Table 9 Analysis of variance table for the response surface quadratic model for R1

Source	Sum of squares	*df*	Mean square	*F*-value	*p*-value (Prob>*F*)	
Model	4.53	9	0.50	8.80	0.0011	Significant
A-molasses	2.58	1	2.58	45.02	0.000053	
B-temperature	0.014	1	0.014	0.24	0.63	
C-incubation time	1.40	1	1.40	24.48	0.00058	
AB	0.16	1	0.16	2.76	0.13	
AC	0.13	1	0.13	2.32	0.16	
BC	0.003	1	0.003	0.05	0.83	
A^2	0.091	1	0.09	1.60	0.24	
B^2	0.031	1	0.03	0.54	0.48	
C^2	0.17	1	0.17	2.91	0.12	
Residual	0.57	10	0.06			
Lack of fit	0.16	5	0.03	0.40	0.83	Not significant
Pure error	0.41	5	0.08			
Cor total	5.12	19				

verified using the Fisher's statistical model (F-value). Table 9 shows the ANOVA for R1 (Ethanol) response. ANOVA of the response surface quadratic model for response R1 had an '*F-value*' of 8.80, which implied that the model was significant. There was only a 0.11% chance that an F-value this large could occur due to noise. The '*p-value*' of the model was 0.00107, which is less than 0.05; this indicates that the model terms are significant and imply that the bioethanol production is sensitive to the factors/coefficients in the model. The factors which have the most significant influence on bioethanol production are molasses (A) and incubation time (C). The '*Lack of Fit F-value*' of 0.40 implies that the lack of fit is not significant relative to the pure error. There is an 83.37% chance that a '*Lack of Fit F-value*' this large could occur due to noise. Non-significant lack of fit is good as we want the model to fit. Signal-to-noise ratio can be measured by another statistical measurement which is known as the '*Adequate precision*'. A ratio greater than 4 is desirable. The ratio of our model is 10.804, which indicates an adequate signal. This model can be therefore used to navigate the design space as well as for further optimization.

'Coefficient of determination or R^2' value gives information about the goodness of fit of a model. It indicates the correlation between experimental and predicted values. If the value of R^2 is closer to 1, it indicates that the filled model explains most of the variability, while a value closer to 0 indicates that there is no linear relationship between values. In the current study, the R^2 value is 0.89, which indicates that the experimental and predicted values are in reasonable agreement. The 'coefficient of

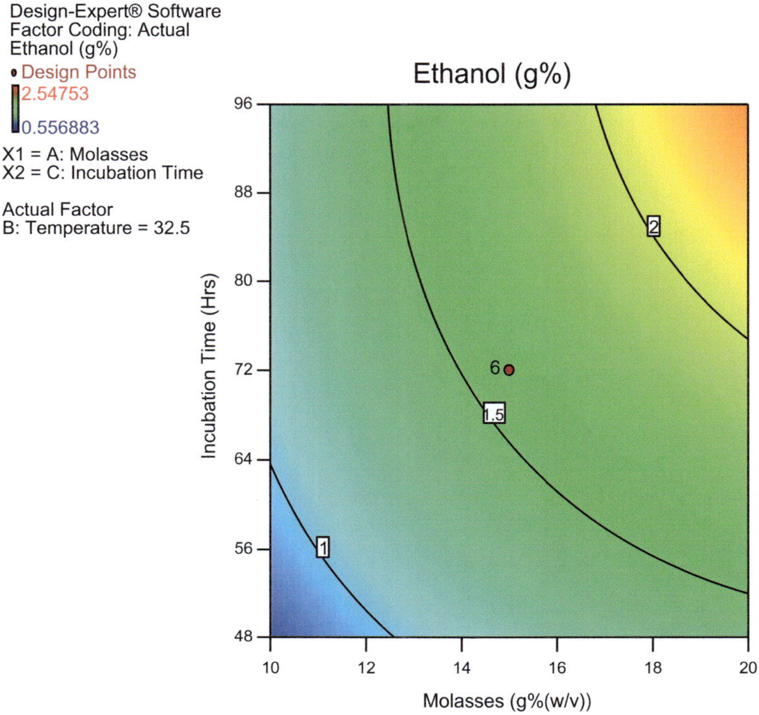

Fig. 7 Contour plot showing cooperative effect of incubation time and molasses on bioethanol production

variation (CV)' is the standardized measure of dispersion; it indicates the degree of precision to which the experiments are compared. The higher reliability of the experiment is usually signified by a high value of CV. In the present study, the CV% value is low (16.67), which implies a good reliability and precision of the experiment.

The 'Predicted coefficient of determination (Pred R^2)' of 0.6398 is in reasonable agreement with the 'Adjusted coefficient of determination (AdjR^2)' of 0.7870, i.e. the difference is less than 0.2. This suggests that the data fits well with the model and gives a decisively good estimate of response for the system. The contour plots below show the interactive effect of incubation time and molasses concentration on bioethanol production when the temperature is 32.5 °C (Fig. 7), and the interactive effect of temperature and molasses concentration on bioethanol production when the incubation time is 72 h (Fig. 8).

From the figures, it can be observed that significantly higher production of bioethanol was obtained with proportional increase in molasses concentration and increase in incubation time (Fig. 7). Substantial production of bioethanol was obtained with increase in molasses concentration and no corresponding increase in temperature (Fig. 8). The perturbation plot (Fig. 9) shows the comparative effects of the three variables on bioethanol production. The sharp curvature of two factors—-

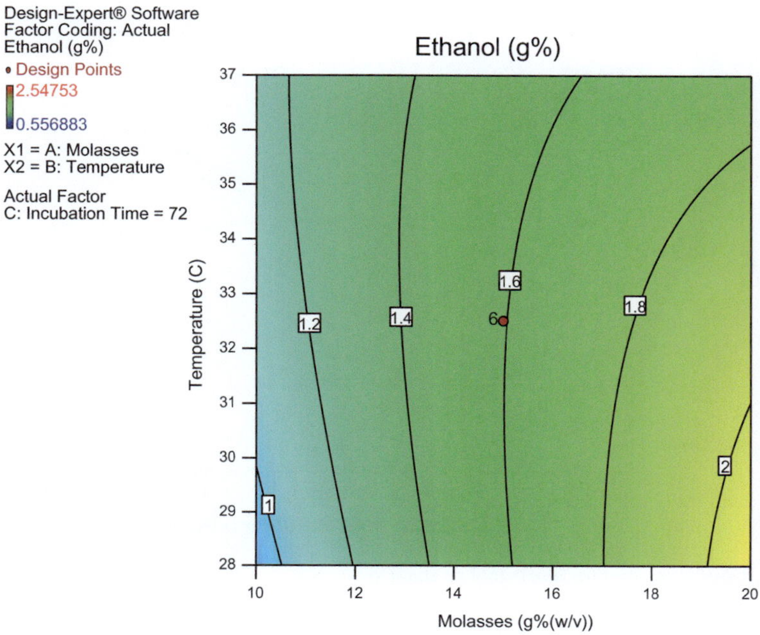

Fig. 8 Contour plot showing cooperative effect of temperature and molasses on bioethanol production

Molasses (A) and incubation time (C)—shows that the bioethanol yield was sensitive to these variables. The comparatively almost flat curve for temperature (B) shows less sensitivity of the response towards this factor. Thus, the temperature of fermentation is not a major variable when immobilized cells are applied for bioethanol production.

The three-dimensional diagram (Fig. 10) displays the interactive effects of molasses concentration and incubation time on bioethanol production at a constant temperature of 32.5 °C. It can be observed from the graph that as the molasses concentration increases, the bioethanol production also increases. The bioethanol concentration also increases when the incubation time is elongated. The simultaneous increase in both the molasses concentration and incubation time shows their cooperative effect on bioethanol production as it increases proportionally to the increase in these two variables. The maximum ethanol production was observed when the molasses concentration was 20 g% (w/v) and the incubation time was 96 h. On the other hand, temperature played no significant role in bioethanol production by immobilized cells of *S. cerevisiae*.

The second response considered is the bioethanol yield (R2). ANOVA of the response surface quadratic model for bioethanol yield is shown in Table 10. The model is a significant model with Fisher F-test value of 3.55, with only a 3.07% chance that an F-value this large could occur due to noise. Values of '*Prob*>*F*' less than 0.0500 indicate that model terms are significant. In this case, C is a significant

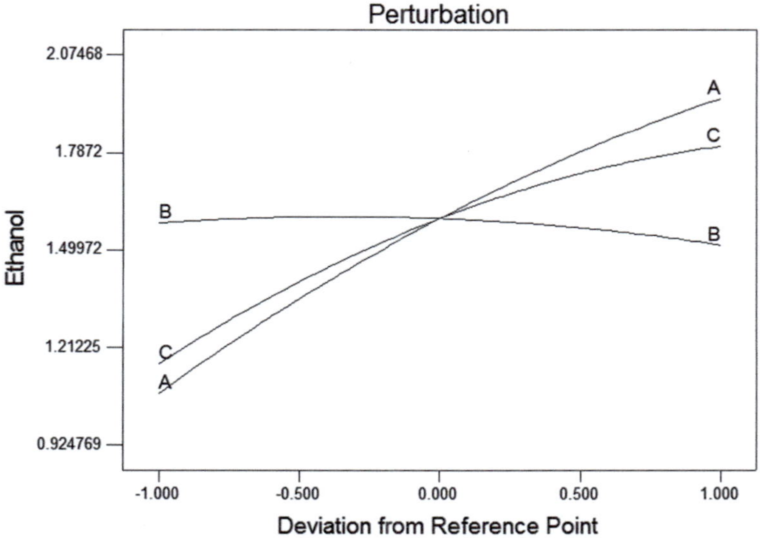

Fig. 9 Perturbation plot for bioethanol production

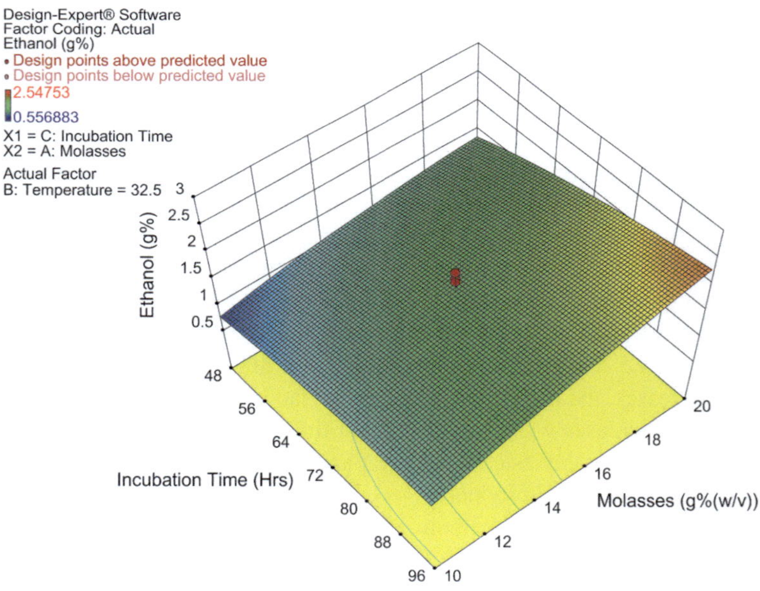

Fig. 10 3D response surface plot showing interaction between molasses and incubation time and their effect on bioethanol production

model term with a '*p-value*' of 0.0005. The goodness of fit for the model was analysed by the value of the coefficient of determination (R^2). In this response, the value of R^2 is 0.76, which indicates that the experimental and predicted values are in reasonable

Table 10 Analysis of variance table for the response surface quadratic model for R2

Source	Sum of squares	df	Mean square	F-value	p-value (Prob>F)	
Model	3408.87	9	378.76	3.55	0.031	Significant
A-molasses	10.10	1	10.10	0.10	0.77	
B-temperature	2.28	1	2.28	0.02	0.89	
C-incubation time	2711.77	1	2711.77	25.38	0.0005	
AB	213.35	1	213.35	2.00	0.19	
AC	4.12	1	4.13	0.04	0.85	
BC	0.89	1	0.89	0.008	0.93	
A2	205.075	1	205.08	1.92	0.20	
B2	44.88	1	44.88	0.42	0.53	
C2	290.05	1	290.05	2.72	0.13	
Residual	1068.38	10	106.838			
Lack of fit	259.16	5	51.83	0.32	0.88	Not significant
Pure error	809.21	5	161.84			
Cor total	4477.25	19				

agreement. The '*Lack of Fit F-value*' of 0.32 implies the lack of fit is not significant relative to the pure error. There is an 88.15% chance that a '*Lack of Fit F-value*' this large could occur due to noise. Non-significant lack of fit is good as we want the model to fit. The CV% value for the model is low at 16.16, which implies a good reliability and precision of the experiment.

The 'Predicted coefficient of determination (Pred R^2)' of 0.6398 is in reasonable agreement with the 'Adjusted coefficient of determination (Adj R^2)' of 0.5466, i.e. the difference is less than 0.2. This suggests that the data fits well with the model and gives a decisively good estimate of response for the system. '*Adequate precision*' measures the signal-to-noise ratio. A ratio greater than 4 is desirable. The ratio of our model is 6.485, which indicates an adequate signal. This model can be therefore be used to navigate the design space as well as for further optimization.

Figure 11 shows a contour plot which shows the effect of incubation time and molasses concentration on bioethanol yield. It can be observed that incubation time is the major variable that affects the yield, while the effect of molasses concentration on the yield is not as significant. From the colour of the graph, it can be deduced that the incubation time between 80 and 96 h is adequate to increase the ethanol concentration to 75% or above. The perturbation plot (Fig. 12) shows the comparative effects of the three variables on bioethanol production. The sharp curvature of one factor, incubation time (C), shows that the bioethanol yield was sensitive to this variable. The comparatively almost flat curve for molasses (A) and temperature (B)

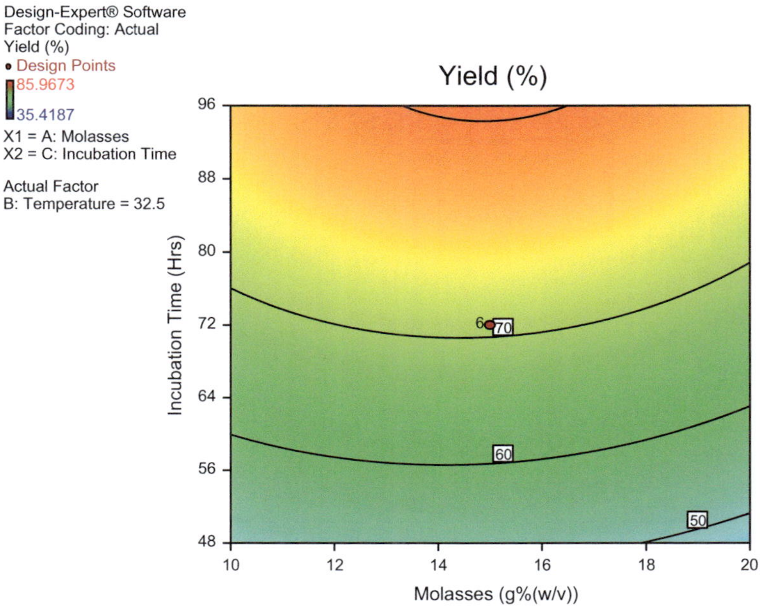

Fig. 11 Contour plot showing cooperative effect of incubation time and molasses on bioethanol yield

showed less sensitivity of the response (i.e. yield) towards those factors. Thus, the molasses concentration and temperature of fermentation are not major variables when bioethanol yield is concerned. The 3D plot (Fig. 13) shows the interactive effects of molasses concentration and incubation time on bioethanol production at a constant temperature of 32.5 °C. It can be observed that as the incubation time increases, the bioethanol yield also increases.

6.3.2 Optimization of Fermentation Process and Model Verification

Statistical methods such as factorial designs and response surface methodologies are widely used for the improvement of several bioproducts, including bioethanol (Joshi et al. 2007; Kshirsagar et al. 2015; Raheem et al. 2015; Turhan et al. 2015). The process of optimization was carried out to determine the optimum value of bioethanol production, using the Design Expert software 9.0.4.1, Stat-Ease, Inc. According to the built-in optimization step, the desired goal for each operational condition, i.e. Molasses (A), temperature (B) and incubation time (C), was chosen within the studied range. The response (bioethanol production) was defined as 'maximum' to achieve the highest performance. The programme combines the individual desirability into a single number and then searches to optimize this function based on the response goal. Accordingly, the optimum working conditions and respective bioethanol production

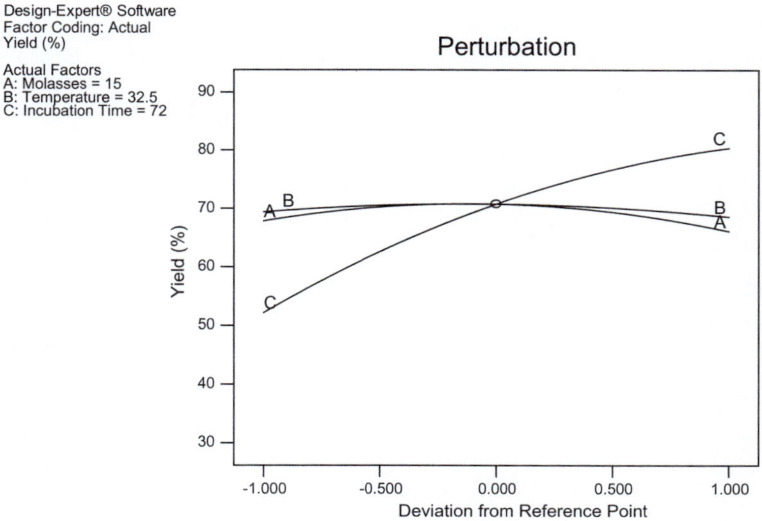

Fig. 12 Perturbation plot for bioethanol yield

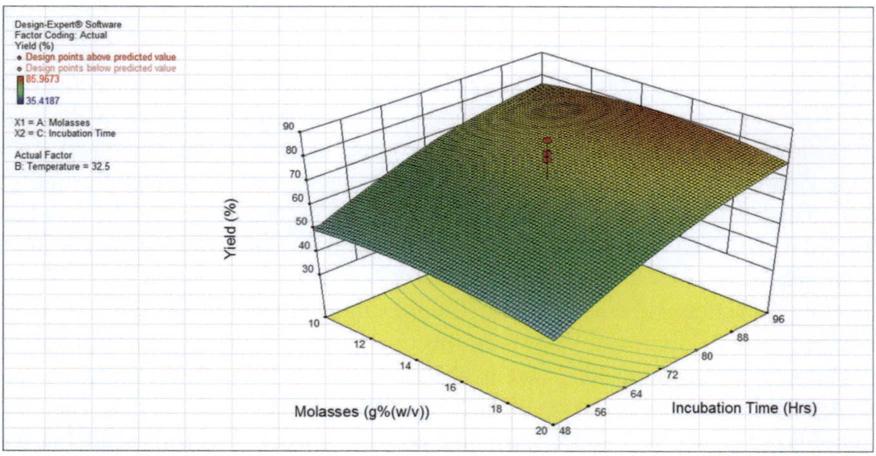

Fig. 13 3D response surface plot showing interaction between molasses and incubation time and their effect on bioethanol yield

were established, and the results are presented in Table 11. The average bioethanol production after optimization was 2.3 g% ± 0.14. Similarly, for the optimization of bioethanol yield, the response was defined as 'maximum' to achieve the highest performance. The optimum working conditions and respective bioethanol yield are presented in Table 12.

The maximum bioethanol yield observed in the study after optimization, which is 91.0245% ± 0.51, was comparable to the theoretical yield in the work done by

Table 11 Optimum condition solutions for bioethanol production

Number	Molasses [g%(w/v)]	Temperature (°C)	Incubation time (h)	Desirability	Ethanol (g%)
1	20	28.00	96.00	0.94	2.44
2	20	28.00	91.09	0.92	2.38
3	20	32.89	96.00	0.86	2.27
4	20	35.76	96.00	0.79	2.13

Table 12 Optimum condition solutions for bioethanol yield

Number	Molasses [g%(w/v)]	Temperature (°C)	Incubation time (h)	Desirability	Ethanol (g%)
1	18.33	28.00	96.00	0.90	91.0245
2	18.50	28.00	95.99	0.90	91.0165
3	18.40	28.00	95.81	0.90	90.9761
4	18.12	29.07	96.00	0.90	90.8148
5	17.14	28.00	96.00	0.88	90.8133
6	13.21	34.22	96.00	0.88	90.1077
7	11.61	36.88	96.00	0.88	89.7512

Göksungur and Zorlu (2001) in Turkey using Ca-alginate immobilized *S. cerevisiae* with beet molasses serving as the substrate. It was also similar to the yield obtained by Ivanova et al. (2011), who obtained an average of 90% of the theoretical yield using CAMB for simultaneous ethanol fermentation and starch saccharification. It was greater than the theoretical yield observed by Limtong et al. (2007), who had utilized *Kluyveromyces marxianus* as the fermentation organism and sugarcane juice as the substrate. The validation of the RSM was carried out to confirm the results of ethanol production and ethanol yield. The maximum ethanol production and yield obtained were 2.75 g% and 91.85%, respectively, at the molasses concentration of 20 g% (w/v), temperature 28 °C and incubation time of 96 h. Due to the incorporation of magnetite in immobilized beads of *S. cerevisiae,* it was observed that when the immobilized beads were added into the production media, the beads settled at the bottom of the conical flask, while the immobilized beads lacking magnetite did not settle to the bottom of the flask and were instead observed to be floating on the surface of the media. Alcohol production is an anaerobic process, so when the beads settle at the bottom of the flask, where there is less oxygen available, they are able to produce alcohol more efficiently.

6.4 Fed-Batch Packed-Bed Fermentation

The ethanol produced and the yield were almost constant in every batch. The average ethanol produced by fed-batch fermentation was 1.832 g% $\pm$0.103. The average ethanol yield was 81.420% ± 4.6. Prakasham et al. (1999) investigated the catalytic role of various inert solid supports on the acceleration of alcoholic fermentation by *S. cerevisiae*. The tested supports were de-lignified sawdust, de-lignified wheat bran, river sand, chitin, chitosan and titanium oxide. The results of the alcoholic fermentation showed that all carriers stimulated ethanol production, which was attributed to the attachment of the cells to these materials. Bekers et al. (1999) used porous spheres of stainless steel treated by oxidation with $TiCl_4$ or aminopropyltrietoxilase, as carriers for yeast cells. The assays of batch fermentation using an inoculum of immobilized cells showed an increase of yeast cell stability and ethanol production. The authors suggested that the increase of ethanol synthesis by cell immobilization in porous treated stainless steel could be the result of catalytic action of some carrier surface element on metabolism. Nigam et al. (1998) carried out alcoholic fermentation using agar-immobilized yeast cells. A packed-bed reactor was employed, and cane molasses was utilized. They obtained a maximum productivity of 79.5 g ethanol/L h with 195 g/L reducing sugar as feed. Low dilution rates are allowed for proper utilization of sugar, which in turn affected the ethanol concentration and volumetric ethanol productivity. The process was continued for 100 days, and the beads remained stable over the course of the fermentation. In the study conducted by Göksungur and Zorlu (2001), it was found that on employing continuous immobilized packed-bed reactor for ethanol production, ethanol concentration of 4.43% and a theoretical yield of 79.5% were observed at the end of 25 days. The 2% calcium alginate beads also retained their structure over the course of fermentation. Osawemwenze and Adogbo (2013) studied the ethanol synthesis using yeast anchored on calcium alginate and clay support. They observed that immobilized yeast cells using clay support gave higher ethanol product yield in both batch and fed-batch processes as compared to calcium alginate support Ivanova et al., (2011). *S. cerevisiae* cells were entrapped in a matrix of alginate and magnetic nanoparticles (CAMB) and covalently immobilized on magnetite-containing chitosan (CHMM) and cellulose-coated magnetic nanoparticles (CCMN). These immobilized cells were applied in column reactors for ethanol fermentation. The type of immobilization affected the ethanol fermentation along with other factors such as feed sugar concentration, initial particle loading and the dilution rate. The overall ethanol yield of 88.8% was obtained using CAMB for ethanol fermentation from starch hydrolysates. Table 13 shows the ethanol production by different microorganisms immobilized on different substrates.

6.5 Evaluation of Calcium Alginate Magnetite Beads (CAMB)

The hardness and rigidity of the CAMB were tested manually by the application of pressure. There was sufficient substrate penetration into the beads due to better porosity, and the beads were strong enough to hold the weight of packing in the column. They were also stable and active for a long time period. The beads could be

Table 13 Immobilization of different microorganisms using a variety of matrices for ethanol production

Type of immobilization matrix	Microorganism	Yield%	References
Calcium alginate	*S. cerevisiae*	97	McGhee et al. (1984)
Calcium alginate	*S. cerevisiae*	95	Nagashima et al. (1983)
Cotton cloth	*S. cerevisiae, K. marxianus, K. fragilis*	90	Joshi and Yamazaki (1984)
Polyurethane	*S. diastaticus*	78.8	Amin et al. (1985)
Radiation polymers	*S. formosensis*	≈40	Fujimura and Kaetsu (1985)
Reticulated polyester foam	*S. cerevisiae, S. uvarum*	98	Black et al. (1984)
Calcium alginate	*Z. mobilis*	95	Bajpai and Margaritis (1985)
Calcium alginate–magnetite	*S. cerevisiae*	91	Present study

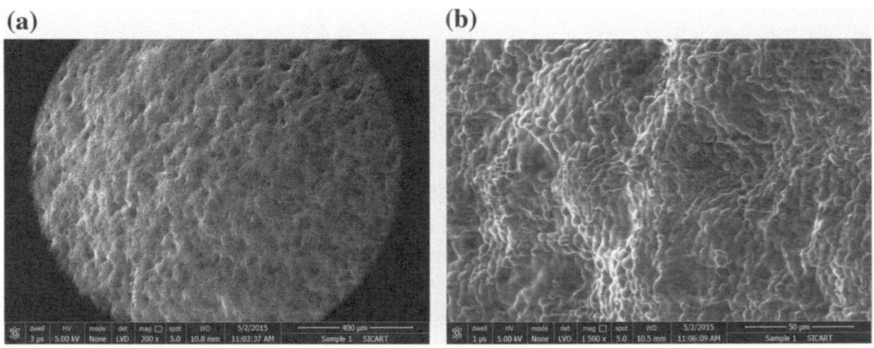

Fig. 14 SEM images of yeast cells immobilized in CAMB, after 96 h fermentation; **a** 200× magnification, and **b** 1500× magnification

stored by refrigeration for more than 100 days. After the CAMB were used for ethanol fermentation, some swelling in the size of the beads was observed. An increase in the size of the beads by almost 10% was observed after repeated ethanol fermentations. The CAMBs with immobilized yeast cells were analysed by ESEM with EDAX to observe the surface structure of the beads. It can be observed that yeast is immobilized in the beads and is actively growing (Fig. 14).

7 Conclusion and Future Outlook

Biofuels are derived from renewable biomass resources, so they are a definite strategic advantage for the promotion of sustainable development of renewable energy resources. They can supplement conventional energy sources in meeting the rapidly increasing requirements for transportation fuels, which can be associated with high economic growth, as well as in meeting the energy needs of any countries vast agrarian, suburban and metropolitan population. To a greater extent, biofuels can satisfy these energy needs in an environmentally benign and cost-effective manner while reducing dependence on import of fossil fuels and thereby providing a higher degree of National Energy Security.

In this study, the effects of multiple factors were evaluated on the basis of different statistical models such as Plackett–Burman and response surface methodology (Central Composite Design). On the basis of Plackett–Burman analysis, it was determined that the model was significant, and the factors of molasses concentration, temperature and incubation time were found significant. Other factors such as potassium di-hydrogen phosphate, ammonium sulphate, magnesium sulphate, yeast extract, pH, immobilized yeast, agitation and pretreated hydrolysate were found to be not significant. This means that all these factors are acceptable at their minimum levels as compared to the significant factors. The reason for this could be that since the cells were immobilized and already in the stationary phase, growth factors and nutrients such as KH_2PO_4, $(NH_4)_2SO_4$, $MgSO_4$ and yeast extract were not required in higher concentration. Since the organisms were immobilized, pH did not adversely affect the rate of bioethanol production. Pretreated hydrolysate, which was added (5%) to observe the effect of furfural compound on *S. cerevisiae*, also did not affect the rate of bioethanol production. On the basis of response surface methodology—central composite design, it was determined that the quadratic model was significant, and the factors of molasses concentration and incubation time were found to be significant. Temperature was not found to be a significant factor. The reason for this could be that since the organisms were immobilized, there was less effect of temperature on the immobilized cells. The immobilized cells could be reused for more than 120 days, retaining its original activity.

Molasses is the current major source for bioethanol production and it is available cheaply due to it being a waste by-product of sugar mills. Molasses is more preferred over lignocellulosic substrates because despite being cheaper than molasses, such substrates require additional treatment before they can be utilized for bioethanol production. Also, molasses has sugars which are readily degraded by microorganisms. When lignocellulosic substrates are given acid treatment, hydroxymethylfurfural (HMF) is produced, which is inhibitory to the production of ethanol by microorganisms.

The main advantage of immobilized system for large-scale industrial production of bioethanol is that it is economically beneficial because it eliminates the need for a separate process of cell removal from the product stream. Also, in this study, the effect of hydroxymethylfurfural (HMF) on the activity of immobilized cells was observed and it was found that immobilized cells could tolerate 5% concentration of HMF. Therefore, immobilized yeast cells can be used for the production of bioethanol from sources which contain diluted sugar such as effluent from paper–pulp industry as well as acid-treated lignocellulose substrates. Further studies using yeast immobilized in CAMB are needed such as continuous fermentation, further scale up and bioethanol production in a magnetically stabilized fluidized bed reactor (MSFBR), where the position of the beads in the system can be controlled and maintained by the application of oscillating electric field.

Acknowledgements SI and VP would like to thank ARIBAS and CVM, and SJ would like to acknowledge Sultan Qaboos University for providing the research facility.

References

Abdul Rahman I, Ayob MTM, Radiman S (2014) Enhanced photocatalytic performance of NiO-decorated ZnO nanowhiskers for methylene blue degradation. J Nanotechnol 212694:8. https://doi.org/10.1155/2014/212694

Al-Bahry SN, Al-Wahaibi YM, Elshafie AE, Al-Bemani AS, Joshi SJ, Al-Makhmari HS, Al-Sulaimani HS (2013) Biosurfactant production by *Bacillus subtilis* B20 using date molasses and its possible application in enhanced oil recovery. Int Biodeterior Biodegradation 81:141–146

Alexandre H, Rousseaux I, Charpentier C (1994) Ethanol adaptation mechanisms in *Saccharomyces cerevisiae*. Biotechnol Appl Biochem 20(2):173–183

Amin G, De Mot R, Van Dijck K, Verachtert H (1985) Direct alcoholic fermentation of starchy biomass using amylolytic yeast strains in batch and immobilized cell systems. Appl Microbiol Biotechnol 22(4):237–245

Bajaj BK, Taank V, Thakur RL (2003) Characterization of yeasts for ethanolic fermentation of molasses with high sugar concentrations. J Sci Ind Res 62(11):1079–1085

Bajaj BK, Yousuf S, Thakur RL (2001) Selection and characterization of yeasts for desirable fermentation characteristics. Indian J Microbiol 41(2):107–110

Bajpai PK, Margaritis A (1985) Kinetics of ethanol production by immobilized cells of *Zymomonas mobilis* at varying D-glucose concentrations. Enzyme Microb Technol 7(9):462–464

Baptista CMSG, Cóias JMA, Oliveira ACM, Oliveira NMC, Rocha JMS, Dempsey MJ, Benson PS (2006) Natural immobilisation of microorganisms for continuous ethanol production. Enzyme Microb Technol 40(1):127–131

Beaven MJ, Charpentier C, Rose AH (1982) Production and tolerance of ethanol in relation to phospholipid fatty-acyl composition in *Saccharomyces cerevisiae* NCYC 431. J Gen Microbiol 128(7):1447–1455

Bekers M, Ventina E, Karsakevich A, Vina I, Rapoport A, Upite D, Linde R (1999) Attachment of yeast to modified stainless steel wire spheres, growth of cells and ethanol production. Process Biochem 35(5):523–530

Berry CC, Curtis AS (2003) Functionalization of magnetic nanoparticles for applications in biomedicine. J Phys D Appl Phys 36(13):R198

Bisson LF (1999) Stuck and sluggish fermentations. Am J Enol Vitic 50(1):107–119

Black GM, Webb C, Matthews TM, Atkinson B (1984) Practical reactor systems for yeast cell immobilization using biomass support particles. Biotechnol Bioeng 26(2):134–141

Borzani W (2001) Variation of the ethanol yield during oscillatory concentrations changes in undisturbed continuous ethanol fermentation of sugar-cane blackstrap molasses. World J Microbiol Biotechnol 17(3):253–258

Borzani W, Gerab A, De La Higuera GA, Pires MH, Piplovic R (1993) Batch ethanol fermentation of molasses: a correlation between the time necessary to complete the fermentation and the initial concentrations of sugar and yeast cells. World J Microbiol Biotechnol 9(2):265–268

Cassidy MB, Lee H, Trevors JT (1996) Environmental applications of immobilized microbial cells: a review. J Ind Microbiol 16(2):79–101

Chen JC, Chou CC (1993) Cane sugar handbook: a manual for cane sugar manufacturers and their Chemists. Wiley, New York

Cohen Y (2001) Biofiltration–the treatment of fluids by microorganisms immobilized into the filter bedding material: a review. Biores Technol 77(3):257–274

De Carvalho JCM, Aquarone E, Sato S, Brazzach ML, Moraes DA, Borzani W (1993) Fed-batch alcoholic fermentation of sugar cane blackstrap molasses: influence of the feeding rate on yeast yield and productivity. Appl Microbiol Biotechnol 38(5):596–598

Del Borghi M, Converti A, Parisi F, Ferraiolo G (1985) Continuous alcohol fermentation in an immobilized cell rotating disk reactor. Biotechnol Bioeng 27(6):761–768

Dubois M, Gilles KA, Amilton JK (1956) Colorimetric determination of sugars and related substances. Anal Chem 28:350–356

El Ghandoor H, Zidan HM, Khalil MM, Ismail MIM (2012) Synthesis and some physical properties of magnetite (Fe_3O_4) nanoparticles. Int J Electrochem Sci 7:5734–5745

El-Gendy NS, Madian HR, Amr SSA (2013) Design and optimization of a process for sugarcane molasses fermentation by *Saccharomyces cerevisiae* using response surface methodology. Int J Microbiol 2013

Fujimura T, Kaetsu I (1985) Nature of yeast-cells immobilized by radiation polymerization activity dependence on the molecular-motion of polymer carriers. Zeitschrift Fur Naturforschung Ca J Biosci 40(7–8):576–579

Ghareib M, Youssef KA, Khalil AA (1988) Ethanol tolerance of *Saccharomyces cerevisiae* and its relationship to lipid content and composition. Folia Microbiol 33(6):447–452

Göksungur Y, Zorlu N (2001) Production of ethanol from beet molasses by Ca-alginate immobilized yeast cells in a packed-bed bioreactor. Turk J Biol 25(3):265–275

Gray WD (1941) Studies on the alcohol tolerance of yeasts. J Bacteriol 42(5):561

Gray WD (1948) Further studies on the alcohol tolerance of yeast: its relationship to cell storage products. J Bacteriol 55(1):53

Gupta AK, Gupta M (2005) Synthesis and surface engineering of iron oxide nanoparticles for biomedical applications. Biomaterials 26(18):3995–4021

Hansen AC, Zhang Q, Lyne PW (2005) Ethanol–diesel fuel blends—a review. Biores Technol 96(3):277–285

Haynes WC, Wickerham LJ, Hesseltine CW (1955) Maintenance of cultures of industrially important microorganisms. Appl Microbiol 3(6):361

Huang SH, Liao MH, Chen DH (2003) Direct binding and characterization of lipase onto magnetic nanoparticles. Biotechnol Prog 19(3):1095–1100

Ibeas JI, Jimenez J (1997) Mitochondrial DNA loss caused by ethanol in *Saccharomyces* flor yeasts. Appl Environ Microbiol 63(1):7–12

Ingale S, Joshi SJ, Gupte A (2014) Production of bioethanol using agricultural waste: banana pseudo stem. Braz J Microbiol 45(3):885–892

Ingram LO (1976) Adaptation of membrane lipids to alcohols. J Bacteriol 125(2):670–678

Ivanova V, Petrova P, Hristov J (2011) Application in the ethanol fermentation of immobilized yeast cells in matrix of alginate/magnetic nanoparticles, on chitosan-magnetite microparticles and cellulose-coated magnetic nanoparticles. Int Rev Chem Eng 3(3):289–299

Joshi S, Yamazaki H (1984) Film fermenter for ethanol production by yeast immobilized on cotton cloth. Biotech Lett 6(12):797–802

Joshi S, Bharucha C, Jha S, Yadav S, Nerurkar A, Desai AJ (2008) Biosurfactant production using molasses and whey under thermophilic conditions. Biores Technol 99(1):195–199

Joshi S, Yadav S, Nerurkar A, Desai AJ (2007) Statistical optimization of medium components for the production of biosurfactant by *Bacillus licheniformis* K51. J Microbiol Biotechnol 17(2):313

Kshirsagar SD, Waghmare PR, Loni PC, Patil SA, Govindwar SP (2015) Dilute acid pretreatment of rice straw, structural characterization and optimization of enzymatic hydrolysis conditions by response surface methodology. RSC Adv 5(58):46525–46533

Limtong S, Sringiew C, Yongmanitchai W (2007) Production of fuel ethanol at high temperature from sugar cane juice by a newly isolated *Kluyveromyces marxianus*. Biores Technol 98(17):3367–3374

López A, Lázaro N, Marqués AM (1997) The interphase technique: a simple method of cell immobilization in gel-beads. J Microbiol Methods 30(3):231–234

Mallick N (2002) Biotechnological potential of immobilized algae for wastewater N, P and metal removal: a review. Biometals 15(4):377–390

Marcelle A, de Vos Betty-Jayne, Visser MS (2007) The preparation, assay and certification of aqueous ethanol reference solutions. Accred Qual Assur 12:188–193

McGhee JE, Carr ME, St Julian G (1984) Continuous bioconversion of starch to ethanol by calcium-alginate immobilized enzymes and yeasts. Cereal Chem (US) 61(5)

Miller GL (1959) Use of DNS reagent for the measurement of reducing sugar. Anal Chem 31(1):426–428

Mino AK (2010) Ethanol production from sugarcane in India: viability, constraints and implications, Doctoral dissertation, University of Illinois at Urbana–Champaign

Mishra P, Prasad R (1989) Relationship between ethanol tolerance and fatty acyl composition of *Saccharomyces cerevisiae*. Appl Microbiol Biotechnol 30(3):294–298

Moreno-Garrido I (2008) Microalgae immobilization: current techniques and uses. Biores Technol 99(10):3949–3964

Muthukrishnan S, Bhakya S, Kumar TS, Rao MV (2015) Biosynthesis, characterization and antibacterial effect of plant-mediated silver nanoparticles using *Ceropegia thwaitesii*–An endemic species. Ind Crops Prod 63:119–124

Nagashima M, Azuma M, Noguchi S (1983) Technology developments in biomass alcohol production in Japan: continuous alcohol production with immobilized microbial cells. Ann NY Acad Sci 413(1):457–468

Najafpour G, Younesi H, Ismail KSK (2004) Ethanol fermentation in an immobilized cell reactor using *Saccharomyces cerevisiae*. Biores Technol 92(3):251–260

Nigam JN, Gogoi BK, Bezbaruah RL (1998) Alcoholic fermentation by agar-immobilized yeast cells. World J Microbiol Biotechnol 14(3):457–459

Nofemele Z, Shukla P, Trussler A, Permaul K, Singh S (2012) Improvement of ethanol production from sugarcane molasses through enhanced nutrient supplementation using *Saccharomyces cerevisiae*. J Brew Distilling 3(2):29–35

Nyirő-Kósa I, Rečnik A, Pósfai M (2012) Novel methods for the synthesis of magnetite nanoparticles with special morphologies and textured assemblages. J Nanopart Res 14(10):1–10

Okita WB, Bonham DB, Gainer JL (1985) Covalent coupling of microorganisms to a cellulosic support. Biotechnol Bioeng 27(5):632–637

Osawemwenze LA, Adogbo GM (2013) Ethanol synthesis using yeast anchored on calcium alginate and clay support. Int J Sci Eng Res 4(4):479–484

Padman AJ, Henderson J, Hodgson S, Rahman PK (2014) Biomediated synthesis of silver nanoparticles using *Exiguobacterium mexicanum*. Biotech Lett 36:2079–2084

Pankhurst QA, Connolly J, Jones SK, Dobson J (2003) Applications of magnetic nanoparticles in biomedicine. J Phys D Appl Phys 36(13):R167

Prakasham RS, Kuriakose B, Ramakrishna SV (1999) The influence of inert solids on ethanol production by *Saccharomyces cerevisiae*. Appl Biochem Biotechnol 82(2):127–134

Pretorius IS (2000) Tailoring wine yeast for the new millennium: novel approaches to the ancient art of winemaking. Yeast 16(8):675–729

Priyadarshini E, Pradhan N, Sukla LB, Panda PK (2014) Controlled synthesis of gold nanoparticles using *Aspergillus terreus* IF0 and its antibacterial potential against gram negative pathogenic bacteria. J Nanotechnol 653198:9. https://doi.org/10.1155/2014/653198

Raheem A, Wakg WA, Yap YT, Danquah MK, Harun R (2015) Optimization of the microalgae *Chlorella vulgaris* for syngas production using central composite design. RSC Adv 5(88):71805–71815

Raju SS, Shinoj P, Joshi PK (2009) Sustainable development of biofuels: prospects and challenges. Econ Polit Wkly 65–72

Ray S, Goldar A, Miglani S (2012) The ethanol blending policy in India. Econ Political Wkly 47(1):23–35

Roy K, Sarkar CK, Ghosh CK (2014) Photocatalytic activity of biogenic silver nanoparticles synthesized using yeast (*Saccharomyces cerevisiae*) extract. Appl Nanosci 1–7

Šafařík I, Šafaříková M (1999) Use of magnetic techniques for the isolation of cells. J Chromatogr B Biomed Sci Appl 722(1):33–53

Šafařík I, Šafaříková M (2002) Magnetic nanoparticles and biosciences. Springer, Vienna, pp 1–23

Šafaříková M, Šafařik I (2001) Immunomagnetic separation of Escherichia coli O26, O111 and O157 from vegetables. Lett Appl Microbiol 33(1):36–39

Singh R, Shedbalkar UU, Wadhwani SA, Chopade BA (2015) Bacteriagenic silver nanoparticles: synthesis, mechanism, and applications. Appl Microbiol Biotechnol 99:4579–4593

Tartaj P, Morales MP, Veintemillas-Verdaguer S, Gonzalez-Carreño T, Serna CJ (2006) Synthesis, properties and biomedical applications of magnetic nanoparticles. Handb Magn Mater 16(5):403–482

Turhan O, Isci A, Mert B, Sakiyan O, Donmez S (2015) Optimization of ethanol production from microfluidized wheat straw by response surface methodology. Prep Biochem Biotechnol 45(8):785–795

Vanaja M, Paulkumar K, Baburaja M, Rajeshkumar S, Gnanajobitha G, Malarkodi C, Sivakavinesan M, Annadurai G (2014) Degradation of methylene blue using biologically synthesized silver nanoparticles. Bioinorg Chem Appl 742346. https://doi.org/10.1155/2014/742346

Wang LH, Hsie MC, Chang CY, Kuo YC, Sang SL, Hsiao HD, Chen HC (1984) Improvement of ethanol productivity from cane molasses by a process using a high yeast cell concentration. ASPAC, Food & Fertilizer Technology Center

Wyman CE, Hinman ND (1990) Ethanol. Appl Biochem Biotechnol 24(1):735–753

Nanotechnology in Biofuels Production: A Novel Approach for Processing and Production of Bioenergy

Anindita Biswas

Abstract Now it's been well understood that only fossil fuel cannot meet our today's fuel need, and we have to have an alternative energy resource to keep our everyday activities going. Reportedly countries like USA, Brazil had tradition of using 'green plant': corn, sugar cane as renewable energy resource rather than non renewable source of energy like 'black fossil fuel', however an industrially applicable practice was required to keep environmental balance. Once researcher had understood the thermal processing of plants cellulosic biomass and lignin, followed by catalytic processing of formed biomass derived compound in liquid phase and catalytic conversion of final products, that practice had minimized health and environmental hazard in compared to when fossil fuel is burnt. Still a more cost effective source of fuel was necessary. Nanomaterials: carbon nano tubes (CNT), graphene, aluminum oxide were the answer. Once enzymes like laccase, lipase are immobilized, these could convert plant cellulose to sugar for repeated cycles of reactions, and produce a fuel without environmental hazards like green house effects.

Keywords Nano-biofuel · CNT · Aluminum oxide · Immobilization · Laccase

1 Introduction

We knew it since the last to last decade only that an alternative source of energy is required to meet the fuel appetite of developing human race. Researchers also were engaged to develop a newer, cheaper, and sustainable fuel resource. Since a long time ethanol has been used as an alternative source of fuel in the United States of America, using corn as 'green plant' source rather using 'black fossil fuel' like gasoline, diesel, or petrol (Balandrin et al. 1985). Whereas sugar cane was counted as a great energy source in Brazil. So it had been understood that, once a potential technology

A. Biswas (✉)
Department of Pharmaceutical Sciences, Faculty of Health Sciences, Sam Higginbottom Institute of Agriculture Technology and Sciences, Allahabad 211007, India
e-mail: i.am.anindita.biswas@gmail.com

N. Srivastava et al. (eds.), *Sustainable Approaches for Biofuels Production Technologies*, Biofuel and Biorefinery Technologies 7,
https://doi.org/10.1007/978-3-319-94797-6_10

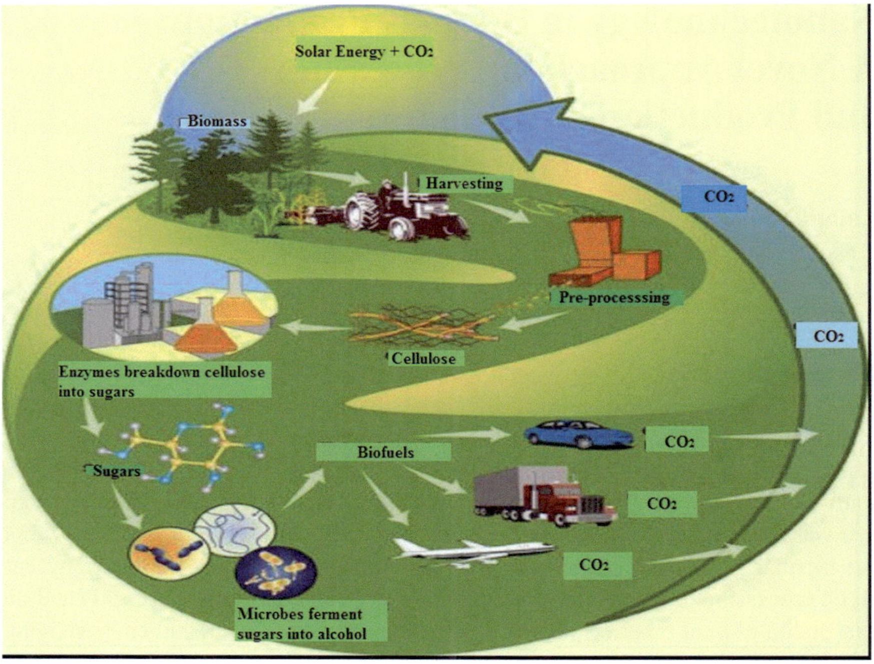

Fig. 1 Cycle of processing–production–usage of biofuel in today's life (Rao 2015)

is found to convert green plant's lignocelluloses (lignin and cellulose) to an economically and environmentally sustainable form of energy (Laborie 2009), that would be easily industrially applicable (Bharathiraja et al. 2014). Research was held mainly on issues like thermal processing of plants cellulosic biomass and lignin, followed by catalytic processing of formed biomass derived compound in liquid phase and finally catalytic conversion of final products and byproducts as well (Bartle and Myers 2001) to minimize health and environmental hazard in compared to produce when fossil fuel is burnt. Modern biotechnology and advanced knowledge of nanotechnology came up with the solution (Whitcombe et al. 2014). Once enzymes required for converting plant cellulose to sugar are immobilized with nanostructured materials (Gao et al. 2014), like multi-walled nanotubes, graphene, the enzymes are industrially used for repeated cycles of reactions, and produce a fuel without environmental hazards like green house effects, in the long run climate change as well. Another potential nanomaterial is aluminum oxide (Siepmann et al. 2008), which plays an important role in platinum/silica catalyst synthesis by strong electrostatic adsorption (SEA). Once in situ real-time tools to monitor catalytic chemistry in atomic scale are developed, like next-generation electron microscope, which would facilitate imaging nanosized chemical changes in aqueous medium, it would lead our footstep towards green environment (Fig. 1).

2 Different Biofuels Construction

Louisiana Tech Professors James Palmer, Yuri Lvov have showed that not only traditional fuel producing plants like sugar cane or corn; woods, grasses, agro wastes, but also household wastes could be considered as fuel source plant, due to their high cellulosic with a savings estimates ranging from approximately $32 million for each cellulosic ethanol plant, because of its reusability.

3 Role of Enzyme in Biofuel Fabrication

Enzymes are used in biofuel production mainly in two stages, one is for hydrolysis of agro waste as pretreatment to produce fermented sugar and another is for trans-esterification while producing biodiesel from plants like Jatropha, algae, or other oil plants. The difficulty with handling of enzyme is their shelf life. Most of the enzymes have half life of few minutes to some hours, as an industrially applicable tool, which has to be increased (Wang et al. 2001) demonstrates that immobilization of enzyme with glass beads or nanostructures could increase that up to thousand times (Fig. 2, Tables 1 and 2).

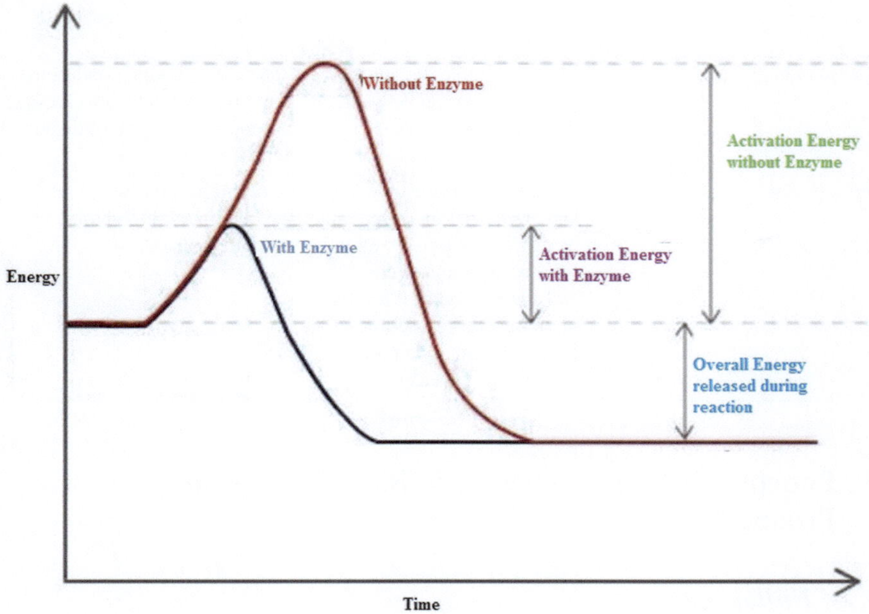

Fig. 2 Activation energy curve with and without enzyme

Table 1 Comparison of free enzyme and immobilized enzymes (Rao 2015)

Characteristics	Immobilized enzyme	Free enzyme
Cost	Low	High
Efficiency	High	Low
Stability	Stable	Unstable
Tolerance to temperature, pH	High	Low
Recovery	Possible	Not possible
Separation from substrate	Easy	Difficult
Separation from product	Easy	Difficult

Table 2 Comparison of different methods to immobilize enzyme (Rao 2015)

Method	Advantages	Disadvantages
Adsorption	Worked in mild condition, easy and low cost, weak interaction between lipase and the carrier make the immobilization, regenerated carrier for several times of usage	Lipase sensitive to pH, ionic strength, and temperature, the adsorption capacity is small and the protein might be stripped off from the carrier
Covalent bond	Thermally and operationally stable enzyme	The laborious preparation of immobilized enzyme might cause lipase to lose its activity, Some coupling reagents are toxic
Cross-linking	Lipase is stable due to the strong interaction between the lipase and the carrier	The cross-linking conditions are intense and the mechanical strength of the immobilized lipase is low
Entrapment	The conditions are moderate and applicable to a wide range of carrier and lipases, effective for low molecular weight substrates because it has the mass transfer restriction during the catalytic process, fast, cheap and easy	Difficulties raise while working with high molecular weight molecule

4 Function of Nanotechnology in Biofuel Processing Production

4.1 Fullerene

Fullerene is a molecule of carbon in the form of a hollow sphere, ellipsoid, tube, or in other shapes. Spherical fullerenes are also called Buckminsterfullerene (Bucky balls).

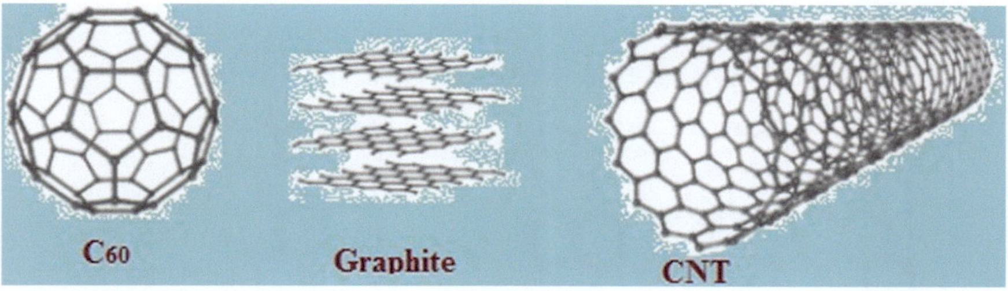

Fig. 3 Depicts different nanoclusters participating in immobilization of enzyme (Rao 2015)

Whereas cylindrical ones are called carbon nanotubes or bucky tubes. Fullerenes are similar in structure to graphite, which is composed of stacked graphene sheets of linked hexagonal rings, but they may also contain pentagonal (sometimes heptagonal) rings. The first fullerene molecule to be discovered, in family is namesake Buckminsterfullerene (C_{60}), was prepared by Richard Smalley, Robert Curl, James Health, Sean O' Brich, and Harold Kroto at Rice University. The name was homage to Buckminster Fuller, whose geodemic domes it resembles (Fig. 3).

4.1.1 Bucky Ball Cluster

Most common type of buckyball is C_{60}. Initially, carbon has only two allotropes, diamond and graphite. Once buckminsterfullerene was discovered, it became the smallest fullerene molecule containing pentagonal and hexagonal rings in which no two pentagons share an edge (which could destabilize the structure, as in pentalene). C_{36}, C_{70}, C_{76}, C_{84} buckyballs are new buckyballs of carbon participating in enzyme immobilization.

4.1.2 Carbon Nanotubes

Carbon atom can form long cylindrical tubes, also known as buckytubes. It is very much possible to make a buckytube with only single atomic layer thick = 1/50,000th that is the thickness of human hair.

Single-Walled Carbon Nanotube

Most single-walled nanotubes (SWNTs) have a diameter of close to one nanometer, and can be conceptualized by wrapping a one-atom-thick layer of graphite called graphene sheet is wrapped and is represented by a pair of indices (n,m). The integers n and m denote the numbers of unit vectors along two directions in the honeycomb crystal lattice of graphene. If $m = 0$, the nanotubes is called zigzag nanotube. If $n = m$, the nanotube is called armchair nanotube, otherwise they are called chiral nanotube.

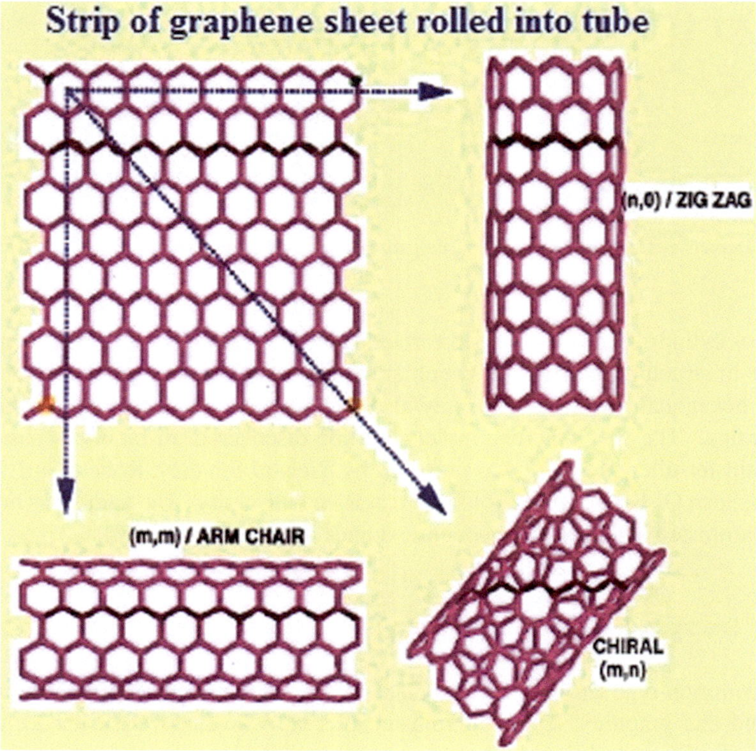

Fig. 4 Different conformations of graphene

Multi-walled Carbon Nanotube

Multi-walled carbon nanotube (MWNTs) consists of multiple rolled layers (concentric tubes) of graphene. These are two models that can be used to describe the structure of multi-walled nanotubes (Fig. 4).

4.2 Aluminum Oxide

Aluminum oxide is another potential nanomaterial used in immobilization of enzymes used in the biofuel production. It participates in the synthesis of platinum/silica catalyst through strong electrostatic adsorption (SEA), and moderates the loading capacity of immobilized enzyme.

5 Blending up Nano-Based Enzyme Models in Biofuel Production

5.1 Reasons Behind Change in Properties Once Immobilized with Nanomaterials

Using nanomaterials in biofuel processing production has twofold advantages. First and foremost is that nanomaterial increases the surface-to-volume ratio (SA:V); hence we end with an enzyme with high loading capacity (Mathew et al. 2009). And the dominance of quantum effects stabilizes the enzyme (Figs. 5 and 6).

5.2 Laccase

Laccase is an external enzyme, produced by various bacteria and fungi. This enzyme is mainly used in second-generation biofuel production (Madhavi and Lele 2009), for pretreatment of agro waste (Pawliszyn 1999), which degrades cellulosic compounds and sugar; and phenolics compounds are formed (Xu et al. 2015). Research has showed immobilization with nanomaterials, such as fullerene (C_{60}), multi-walled carbon nanotubes (MWNTs), oxidized-MWNTs (O-MWNTs), and graphene oxide

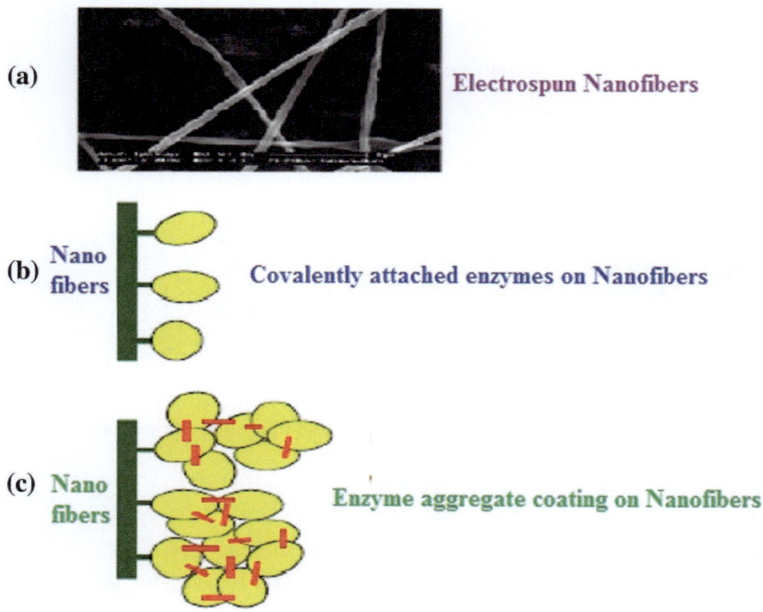

Fig. 5 Different techniques of enzyme immobilization; **a** Electrospun method, **b** Covalently attached, **c** Enzyme aggregate coatings (Rao 2015)

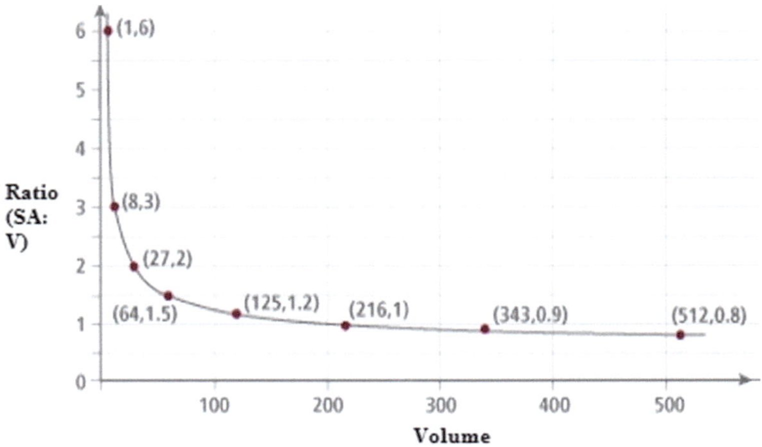

Fig. 6 Curve depicts the change of SA:V with change in volume, after conjugation with nanomaterials

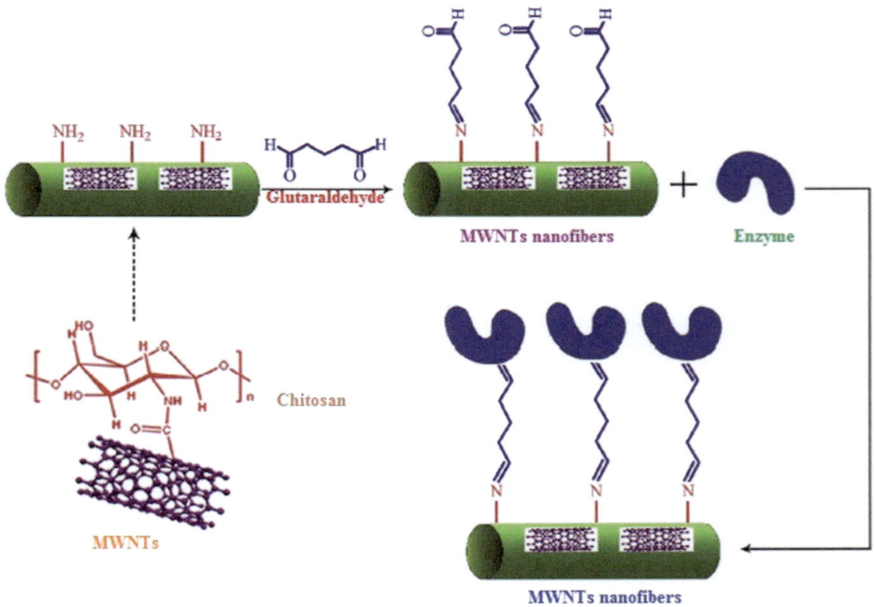

Fig. 7 Schematic demonstration of laccase enzyme immobilization via activation on MWNTs nanofiber membrane (Rao 2015)

(GO) increases the half life of laccase and stabilize it as well (Yücel et al. 2012). Here, O-MWNTs show the maximum loading capacity for enzyme laccase, whereas C_{60} demonstrates the lowest (Fig. 7).

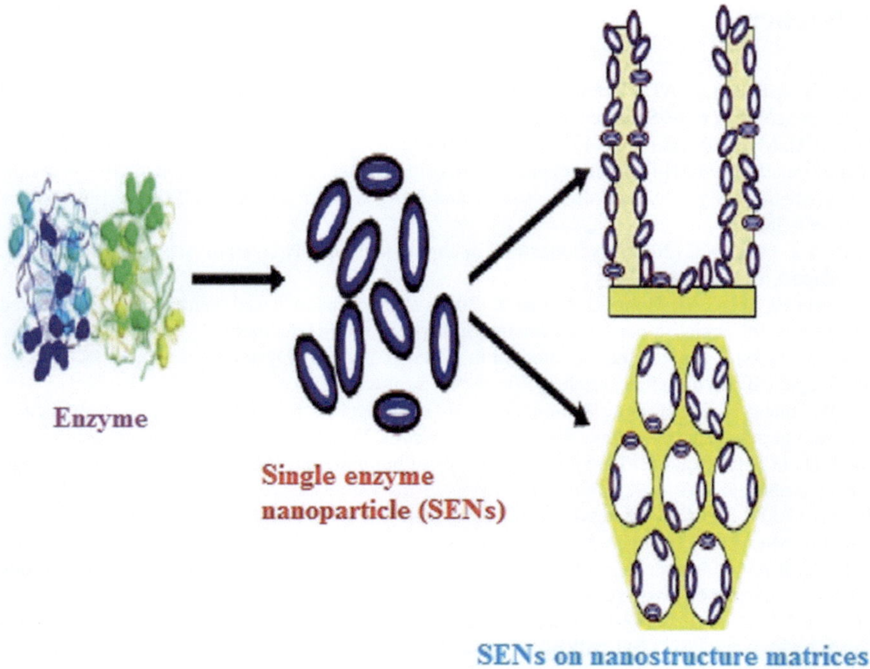

Fig. 8 Immobilization of SENs

5.3 *Lipase*

Biodiesel is usually known as methyl (or ethyl) esters of fatty acids obtained by trans-esterification (alcoholysis) of triglycerides. Lipase an extracellular or intracellular enzyme, obtained from fungi are immobilized in biomass support particles and used as catalytic beds to obtain prolong use (Yücel et al. 2012) (Fig. 8).

6 Conclusion

Nanotechnology plays a significant role in biofuel production, by assisting the immobilization of enzymes like laccase or lipase. As nanostructured enzyme has high surface area compared to free enzymes, immobilized enzymes have increased the shelf life or reusability up to thousand times. Though there is a limitation of enzymes with inhibitors, nanostructured immobilized enzyme shows equivalent catalytic activity when compared to free enzyme.

References

Balandrin MF, Kjocke AJ, Wurtele E et al (1985) Natural plant chemicals: sources of industrial and mechanical materials. Science 228:1154–1160

Bartle KD, Myers P (eds) (2001) Capillary electrochromatography. University of Leeds, UK

Bharathiraja B et al (2014) Biodiesel production using chemical and biological methods—a review of process, catalyst, acyl acceptor, source and process variables. Renew Sustain Energy Rev 38:368–382

Cserháti T, Forgács E (2003) Cyclodextrins in chromatography. Hungarian Academy of Sciences, Budapest, Hungary

Cywinski PJ, Moro AJ, Mohr GJ, Kutner K (2010) Chemosensors based on molecularly imprinted polymers. In: Topics in current chemistry. Springer, Berlin–Heidelberg

Dyson N (1998) Chromatographic integration methods, 2nd edn. Dyson Instruments, UK

Future applications of nanotechnology, Trade News Singapore

Gao W, Huang H, Shi H, Feng X, Song W (2014). Nitrogen-rich graphene from small molecules as high performance anode material. *Nanotechnology* 25:415402

Guo F et al (2012) Solid acid mediated hydrolysis of biomass for producing biofuels. Prog Energy and Combust Sci 38(5):672–690

Hanai T (2000) HPLC: a practical guide. Health Research Foundation, Kyoto, Japan

Hanna R, Maciej JN, Lapinsky L, Adamowicz L (1998) Molecular structure and infra red spectra of 2-hydroxy-1, 4-naphthaquinone; Experimental matrix isolation and theoretical Hartree-Fock and post Hartree-Fock study. Spec Act 54: 1091–1103

Johnston CT (2010) Probing the nanoscale architecture of clay minerals. Clay Miner 45:245–279

Kim J, Jia H, Wang P (2006) Challenges in biocatalysis for enzyme-based biofuel cells. Biotechnol Adv 24(3):296–308

Laborie MP (2009) Chapter 9. Bacterial cellulose and its polymeric nanocomposites. In: Lucia L, Rojas O (eds) The nanoscience and technology of renewable biomaterials. Blackwell Publishing, Hoboken

Lapshin RV (2011) Feature-oriented scanning probe microscopy. In: Nalwa HS (ed) Encyclopedia of Nanoscience and Nanotechnology. American Scientific Publishers, USA, pp 105–115

Madhavi V, Lele SS (2009) Laccase: properties and applications. BioResources 4(4):1694–1717

Mathew AP, Laborie M-P, Oksman K (2009) Cross-linked chitosan-chitin whiskers nanocomposites with improved permeation selectivity and pH stability. Biomacromolecules 10(6):1627–1632

Mehta CM (2012) Role of microbiologically rich compost in reducing Biotic and abiotic stresses. Microorganisms in Environmental Management

Mukhopadhyay A, Dasgupta AK, Chakrabarti K (2014) Enhanced functionality and stabilization of a cold active laccase using nanotechnology based activation-immobilization. Bioresource Technology

Noraini MY et al (2014) A review on potential enzymatic reaction for biofuel production from algae. Renew Sustain Energy Rev 39:24–34

Pang R, Li M, Zhang C (2015) Degradation of phenolic compounds by laccase immobilized on carbon nanomaterials: diffusional limitation investigation. Talanta 131:38–45

Pavlidis IV et al (2012) Development of effective nanobiocatalytic systems through the immobilization of hydrolases on functionalized carbon-based nanomaterials. Bioresour Technol 115: 164–171

Pawliszyn J (ed) (1999) Applications of solid phase microextraction. University of Waterloo, Waterloo, Ontario, Canada

Putzbach W, Ronkainen NJ (2013) Immobilization techniques in the fabrication of nanomaterial-based electrochemical biosensors: a review. Sensors 13(4):4811–4840

Rao A (2015) Nanotechnology: Applications in biofuel productions. Conference paper, VIT University, January 2015 (https://www.reserachgate.net/publication/271902586)

Shen Y (2014) Carbon dioxide bio-fixation and wastewater treatment via algae photochemical synthesis for biofuels production. RSC Adv 4(91):49672–49722

Siepmann F, Siepmann J, Walther M, MacRae RJ, Bodmeier R (2008) Polymer blends for controlled release coatings. J Controlled Release 125:1–15

Szpunar J, Lobinski R (eds) (2003) Hyphenated techniques in speciation analysis. CNRS, Pau, France

Whitcombe MJ, Kirsch N, Nicholls IA (2014) Molecular imprinting science and technology: a survey of the literature for the years from 2004 to 2011. J Mol 27:297–401

Xu R et al (2015) Enhancement of catalytic activity of immobilized laccase for diclofenac biodegradation by carbon nanotubes. Chem Eng J 262:88–95

Yücel S, Terzioğlu P, Özçimen D (2012) Lipase applications in biodiesel production

MIX
Papier aus verantwortungsvollen Quellen
Paper from responsible sources
FSC® C105338

If you have any concerns about our products,
you can contact us on
ProductSafety@springernature.com

In case Publisher is established outside the EU,
the EU authorized representative is:
Springer Nature Customer Service Center GmbH
Europaplatz 3, 69115 Heidelberg, Germany

Printed by Libri Plureos GmbH
in Hamburg, Germany